Indian Fights & Fighters

THE LAST OF CUSTER

Drawing by E. L. Blumenschein

Indian Fights and Fighters

CYRUS TOWNSEND BRADY, LL.D.

Introduction by JAMES T. KING

UNIVERSITY OF NEBRASKA PRESS • LINCOLN/LONDON

To that most eminent and useful

churchman and citizen

Ozi William Whitaker, D.D., LL.D.

Bishop of Pennsylvania

whom I admire as a cleric, respect as a

man, and love as a friend

I dedicate

this story of the West

he served so well

International Standard Book Number 0–8032–5743–0 pbk.
International Standard Book Number 0–8032–1152–X
Library of Congress Catalog Card Number 74–156373

First Bison Book printing: November 1971

Most recent printing shown by first digit below:

8 9 10

The text of the Bison Book edition is reproduced from the first edition
published in 1904 by McClure, Philips & Co.

Manufactured in the United States of America

INTRODUCTION

ALTHOUGH Cyrus Townsend Brady's *Indian Fights and Fighters* (1904) regularly is listed in bibliographies of significant works on the history of the American West, it has long been out of print and copies have become increasingly difficult to obtain. Embracing almost a quarter century in the history of the struggle for the Great Plains, it fully deserves reissue in a form which will make it once again widely available, for it includes not only Brady's own clear, fast-paced narratives of the Plains wars, but also a number of eye-witness accounts, most of which were written especially for Brady and which are almost impossible to find elsewhere.

Cyrus Townsend Brady was born in Allegheny, Pennsylvania, in 1861, the son of a well-to-do banker and accountant, Jasper E. Brady, and his wife, Harriet Townsend Brady. The Brady forebears were a hardy breed of Scotch-Irish Presbyterians; in the eighteenth century their axes and muskets had helped to subdue the Pennsylvania wilderness, giving the family a strong tradition of frontier belligerence. It appeared that Brady would be following the example of his martial ancestors when he graduated from the United States Naval Academy at Annapolis in 1883, but, for reasons which are not clear, he found three years in the navy to be enough. In 1886, with his wife, the former Clarissa

Sidney Guthrie, and their young family, he moved west to Omaha, Nebraska, where he worked for the Missouri Pacific and Union Pacific railroads.

Brady's time in Omaha, though brief, was a turning point in his life. Soon after becoming acquainted with the dean of the Episcopal cathedral, he joined the church and decided to study for the priesthood under Bishop George Worthington. Since he had a family to support, Brady had to continue working, and he recalled later that he read theology during his lunch hour, on the trolley, and until late at night. He was ordained a deacon in 1889 and a priest in 1890, a year marred by the death of his wife. After a short period as a member of the cathedral clergy, on impulse Brady volunteered for missionary service. Energetic, large-hearted, and self-confident, he nevertheless found missionary work exhausting. His responsibilities carried him into five western states, with tours of duty as rector of several small churches in Missouri and Colorado, and as archdeacon of Kansas. "In three years, by actual count," he wrote in *Recollections of a Missionary in the Great West*, "I travelled over 91,000 miles by railroad, wagon and on horseback, preaching or delivering addresses upward of 11,000 times, besides writing letters, papers, making calls, marrying, baptizing, and doing all the other endless work of an itinerant missionary."[1]

In 1895 Brady returned to the East, where he served until 1899 as archdeacon of Pennsylvania and as rector of St. Paul's Church in Philadelphia until 1902. During

1. Brady's *Recollections of a Missionary in the Great West* (New York: Charles Scribner's Sons, 1900) is an anecdotal account of his missionary service; although it reveals his impressions of the West, it contains little biographical material. Among the few published accounts of Brady's life are the brief sketch in the *Dictionary of American Biography* and his obituary in the *New York Times*, January 25, 1920.

the Spanish-American War he was chaplain of the First Pennsylvania Volunteer Infantry. He deplored what he believed to be eastern ignorance about the West—"I have found it necessary," he observed, "to inform the curious that we did not live in tepees or wigwams when in Nebraska or Colorado"—and he celebrated "the advantages presented by life in the West for the higher development of character." In comparison with the westerner, the easterner came off a clear second-best. "Western people," he was convinced, "are usually brighter, quicker, more progressive, and less conservative, and more liberal" than their eastern cousins. "The survival of the fittest is the rule out there," he asserted, striking a note which might have pleased both Theodore Roosevelt and Herbert Spencer, "and the qualities of character necessary to that end are brought to the top by the strenuous life necessitated by the hardships of the frontier."

In 1898, after his return to Philadelphia, Brady published *For Love of Country,* a novel of the American Revolution. It was the first of a long series of romantic novels, among them *When Blades Are Out and Love's Afield, A Little Traitor to the South, Bob Dashaway in Frozen Seas, Little Angel of Canyon Creek, The Eagle of the Empire,* and *A Baby of the Frontier.* Although widely read in their day, Brady's novels were thin, unimaginative, and cloyingly sentimental, and are now forgotten.

As his literary work began to make greater demands on his time, Brady resigned his post at St. Paul's and from 1902 to 1905 devoted himself to writing. *Indian Fights and Fighters* was one of several histories he produced during this period, which also saw the publication of at least ten more novels. Prior to book publication many were serialized in such popular magazines as *Munsey's, Pearson's,* and the *Ladies' Home Journal,* thus augmenting his profits. Yet the substantial income from his writing seems to have been insufficient to provide for a growing family. Brady had married again—his second wife was the former Mary Barrett—and ultimately he had six children, three sons and three daughters. He continued to turn out books, but in 1905 he returned to active church work as rector of Trinity Church in Toledo, Ohio. He resigned from Trinity in 1909—"because," according to one account, "the church was specially incorporated under state laws and not under the canon laws"[2]—and moved to St. George's Church in Kansas City, Missouri, where he remained until 1913. He spent the following year at the Church of the Ascension in Mount Vernon, New York, resigning because he considered the salary inadequate. From that time on, while he maintained an ecclesiastical connection through occasional services for St. Stephen's Church in New York City, he concentrated primarily upon his writing, producing three or four books a year. Eventually he also wrote motion picture scenarios. Others recognized the potential value of Brady's stories to the screen: the publisher Frank A. Munsey, who had bought one for $240, promptly resold it at a handsome profit to a film company. Brady went to court and was awarded two thousand dollars.

Of Brady's vast literary output, only his histories have stood the test of time, and of the histories *Indian Fights and Fighters* is the best. Its strength comes partly

2. *New York Times,* January 25, 1920.

from Brady's knowledge of the Great Plains and from his acquaintance with some of the veterans of the Plains wars (in Omaha, for example, he became a friend of Guy V. Henry, whom he eulogizes in Chapter Twelve), but even more from his determination that the perceptions and recollections of participants in the Indian wars should be preserved. He was tireless in his search for source material. He himself wrote well and when he reprinted verbatim the narratives of others his introductions and annotations were careful and generally satisfactory. It is these narratives, as Brady was well aware, which contribute most to the book's timeless quality.

By modern standards the Indian-fighting army which Brady chronicled was a remarkably small establishment. Reduced to approximately thirty thousand men shortly after the Civil war, it remained about that size, or even smaller, until the outbreak of the Spanish-American War. Part of the post-Civil War Army was stationed in the East—during the Reconstruction years a sizable portion of it was in the South—and the rest was scattered thinly throughout the frontier. Promotion was painfully slow, and ordinarily came only upon the death or retirement of a senior officer. More rapid advancement sometimes could be achieved through the machinations of army politics—a murky tangle of jealousies, rivalries, and dissensions—but most officers resigned themselves to long years of service in grade, and it was not unknown for a thirty-year veteran to retire as a lieutenant or captain.

Frontier duty was hard. The wilderness surrounding an isolated post could often be as hostile as the Indian who battled to keep the white intruders out of his homeland. The Dakotas' arctic cold might be exchanged for the desert's paralyzing heat as a regiment shifted its station from one to another department. C. R. Hauser, the sergeant major of the Fifth Cavalry, once observed that service in such extremes of weather—often without proper shelter or supplies—"made old men of young men, and carried away many a hardy soul, when he should have been in his prime."[3] One of the Indian's best defenses was the terrain, and his cleverness in its use equaled his bravery and skill as a warrior. The frontier soldier, therefore, regularly found himself tramping across endless plains, barren deserts, or rugged mountains, his parade-ground blues replaced by faded denim or ragged corduroy, subsisting on a diet of antique salt pork and tooth-defying hardtack.

The army had other problems as well. Among them were a desertion rate almost as high as that during the Civil War, lack of training, and a chronic shortage of equipment and supplies. These difficulties were overcome in part by the strong sense of duty in that great majority of men who chose not to desert, and by the stern efficiency of a corps of noncommissioned officers whose years of service often stretched from the Civil War to the "Ghost Dance War" of 1890–1891. The regular army was also blessed with a remarkably able officer corps. The "Benzine Boards" after Appomattox had "dry-cleaned" the army of many incompetent and inefficient officers, and those who remained would dominate it until the end of the Indian wars. With some notable exceptions, frontier officers as a rule were intelligent, well educated and well trained. A surprisingly large number were authors, poets, and scholars. A still

3. Quoted in Don Rickey's excellent account of the enlisted man in the frontier Army, *Forty Miles a Day on Beans and Hay* (Norman: University of Oklahoma Press, 1963), p. 254.

larger number were voracious readers, and in many campaign saddlebags, volumes of Shakespeare, Sir Walter Scott, or Thucydides (in the original Greek) might share space with hardtack and coffee.

According to the stereotype, the frontier soldier was supposed to feel that the only good Indian was a dead one. The few surviving memoirs and letters of enlisted men suggest that they, at least, tended to accept this viewpoint, and General William Tecumseh Sherman once remarked that killing off the Indians was the only way to prevent their becoming government-supported paupers. Yet a goodly number of regular officers who served in the field were strongly sympathetic toward the enemy they were sent to fight. Stationed on the fringes of settlement, the frontier officer had first-hand knowledge of treaties broken by the white man, of the cupidity and venality of many Indian agents and traders, and of the desperation of the Indian as the tide of white civilization swept mindlessly into his hunting grounds. General George Crook, for one, put the matter quite frankly. The white man, he observed in 1878, had "come and occupied about all the lands the Indians derived their living from. . . . The disappearance of the game, which means starvation, may seem a small thing to us, but to them it is their all, and he must be a very contemptible being who would not fight for his life."[4] Although their solutions to the "Indian problem" varied widely, the officers of the regular army seldom felt that waging war against the Indians was the answer.

4. Omaha (Nebraska) *Herald*, July 28, 1878. For an interesting discussion of this theme, see Richard N. Ellis, "The Humanitarian Soldiers," *Journal of Arizona History*, X (Summer, 1969). Even Sherman's attitude was usually more moderate than his comment would indicate.

Such views seem to have influenced Washington policy-makers rarely, if at all.

Another characteristic of the frontier army was its cosmopolitan composition. In 1869, the company commanders of the Fifth Cavalry, for instance, included three Irishmen, a Prussian, two Englishmen, and a German.[5] George Custer's Seventh Cavalry officers included, besides three West Pointers, "a Frenchman, a Prussian, a former member of Congress, a mixed-blood Indian, a former judge, a former papal Zouave, and a grandson of Alexander Hamilton."[6] This pattern was repeated in the ranks, where one might hear the soft drawl of an ex-Confederate mixing easily with the gutteral accents of a German or the brogue of an Irishman. Other characteristics of the enlisted men varied as widely as their national backgrounds. Reasons for enlisting ranged from financial problems or evasion of punishment by the law to the simple desire to be a soldier. Don Rickey has observed that a typical group of recruits might include "a mechanic or factory worker tired of dull routine, a farmer with a hankering for frontier life, a restless student tired of his books, a bankrupt businessman, a few shiftless drifters, and some emigrants who could not find work."[7] Many found the escape or glory they sought; some found an early grave.

Such was the army which in the post–Civil War period became a means by which the American vision of Progress was fulfilled on the Great Plains and throughout the West. "The rights of savagery," as Cyrus Town-

5. George F. Price, *Across the Continent with the Fifth Cavalry* (New York: D. Van Nostrand, 1883), pp. 391–420.

6. Oliver Knight, *Following the Indian Wars* (Norman: University of Oklahoma Press, 1960), p. 22.

7. Rickey, *Forty Miles a Day*, p. 21.

send Brady puts it in his first chapter, were gradually "compelled to yield to the demands of civilization, ethics to the contrary notwithstanding." *Indian Fights and Fighters* illustrates one phase of that struggle.

Any book which relies heavily upon recollections of events long past is likely to be somewhat uneven, and such is the case with *Indian Fights and Fighters.* But even its flaws provide unusual insights into the history of the American West. Chapter Six in Part Two is a case in point. In his eagerness to present participants' accounts, Brady reprinted the story told by Theodore Goldin, a prominent Wisconsin Republican, who claimed to have been one of the last men to see Custer alive. Colonel W. A. Graham, however, has shown in *The Custer Myth* that Goldin's story was almost completely a fabrication and that, although Goldin did indeed serve with the Seventh Cavalry, he served neither as long as nor in the capacity he claimed. Goldin's motives appear to have been harmless enough, but in his yearning to have played a larger (though not crucial) part in the historic episode, he added his own bit to the large collection of apochrypha about the battle of the Little Big Horn. Apart from the Goldin account, *Indian Fights and Fighters* still holds much for those who are interested in the Custer catastrophe. Colonel Graham has correctly observed, for instance, that Brady's long appendix contains "a discussion which every student of that disastrous event should read and ponder well."[8]

A problem of a somewhat different nature arose from the exchange between General Louis H. Carpenter

8. W. A. Graham, *The Custer Myth* (Harrisburg, Pa.: Stackpole, 1953), p. 267. Chapter VI is an analysis of Goldin's various accounts, including the letter to Brady.

and General Eugene A. Carr in Chapters Eight and Nine, in Part One. It was not unusual, in the years following the end of the Indian wars, for disputes to arise among the grizzled veterans of those conflicts; once retired, the old soldiers had the leisure to fight their battles once again in their memories, to correct real or fancied slights and affronts, and to defend their records. Few of the veterans' disputes, however, were ever presented so directly before the public as was the one between Carpenter and Carr.

In keeping with Brady's favorite practice, the articles which would become *Indian Fights and Fighters* were first published in *Pearson's Magazine* in 1904. Brady had asked General Carpenter for his recollections of the fight at Beaver Creek in 1868, and Carpenter's reply was published with Brady's annotations. The article was read by General Carr, by then retired from the army for over a decade, who at the time of the battle was being escorted to his regiment by Carpenter's Tenth Cavalry command. Convinced that Carpenter was claiming much credit that should have been his, and angered by Brady's characterization of him as "a passive spectator" during the battle, Carr responded with a long letter to the editor of *Pearson's,* most of which appears in the book as a part of Chapter Nine. Brady, who was sincerely sorry for having contributed to this wrangle between two of the Old Army's finest, tried to patch things up with an apology in the next issue of *Pearson's* and by promising to amend the offending notes when the book was published. But he left the clear implication that he favored Carpenter's interpretation of the events. When he later turned to the battle of Summit Springs, in which Carr had defeated Tall Bull's Dog

Soldiers in 1869, Brady attempted to obtain Carr's recollections of the battle, but the old general, still deeply offended, refused even to reply to Brady's letters. As a result, the battle of Summit Springs is reported in this book by one J. E. Welch, whose recollections are demonstrably inaccurate at many points: for example, Tall Bull, who was a distinguished Cheyenne war leader and chief of the warrior society known as the Dog Soldiers, is wrongly identified as a Sioux. Further, Welch places Tall Bull's death at the hands of a "Lieutenant Mason"– evidently Lieut. George F. Mason of the Fifth Cavalry–while the evidence indicates that the fatal bullet was fired either by the regimental chief of scouts, William F. Cody, or–more likely–by Major Frank North, commander of the famous Battalion of Pawnee Scouts. Welch also has exaggerated the number of Indian dead by twice and the number of wounded by more than seven times. The battle of Summit Springs deserved more satisfactory treatment than Brady was able to accord it; a disaster for the Cheyennes, it was the last significant Indian conflict with the white man on the Central Plains.[9] Welch's recollections, hazy as they are, should be read with caution.

Another feature of this book which deserves comment is the section dealing with Fort Phil Kearny and the so-called Fetterman Massacre of 1866. As Brady notes in his preface, the section was "read and corrected" by Colonel Henry B. Carrington, who was post commandant when William J. Fetterman–according

9. The Battle of Summit Springs is discussed from different viewpoints in Donald F. Danker, ed., *Man of the Plains: Recollections of Luther North* (Lincoln: University of Nebraska Press, 1961), James T. King, *War Eagle: A Life of General Eugene A. Carr* (Lincoln: University of Nebraska Press, 1963), and Don Russell, *The Lives and Legends of Buffalo Bill* (Norman: University *of Oklahoma Press,* 1960).

to Carrington, at least—brashly disobeyed orders and fell into an Indian ambush. Although historians have tended to accept Carrington's version of the incident, J. W. Vaughn has argued recently that it may in fact have been Carrington's orders which sent Fetterman to his death and that the post commander's story was perhaps an attempt to shift blame from himself.[10] Nevertheless, Brady's presentation, by any interpretation, remains a clear and cogent statement of Carrington's point of view.

Lastly, while Brady was undoubtedly sincere in his attempt to treat the Indian fairly, this is a white man's history of the Indian wars. Ironically, even the "Personal Story of Rain-In-The-Face" is presented by a white man and strictly on the white man's terms. Although he was a contemporary of George Bird Grinnell, whose classic *The Fighting Cheyennes* presents the Indian viewpoint with sympathy and clarity, Brady, like most other historians of his generation, lacked both the inclination and the ability of a Grinnell to search out and to use Indian sources. But men like Grinnell are rare in any generation, and Brady did not pretend to be doing anything other than what he did—to present as fairly as he could the stories of white men at war with the Indians of the Great Plains. Considering his gift for swift-moving narrative, he was far too modest in suggesting that he may have told those stories "indifferently"; he was closer to the truth when he added that "the stories at least are there. They speak for themselves. I could not spoil them if I tried."

10. J. W. Vaughn, *Indian Fights: New Facts on Seven Encounters* (Norman: University of Oklahoma Press, 1966), pp. 14–90.

Late in January, 1920, Brady contracted a heavy cold which turned into pneumonia, and after only two days' illness he died. He was not yet sixty years of age. In more than twenty years of writing, he had produced much which spoke only to his own time. But the place of *Indian Fights and Fighters* in the historiography of the American West is secure. In this book, Cyrus Townsend Brady perhaps best fulfilled his hope that in "thus striving to preserve the records of those stirring times I have done history and posterity a service."

JAMES T. KING

University of Wisconsin
River Falls

CONTENTS

Part I

PROTECTING THE FRONTIER

Contents

Part II

THE WAR WITH THE SIOUX

APPENDICES

Part I

Protecting the Frontier

CHAPTER ONE

The Powder River Expedition

I. The Field and the Fighters

SINCE the United States began to be there never was such a post as Fort Philip Kearney, commonly called Fort Phil Kearney.* From its establishment, in 1866, to its abandonment, some two years later, it was practically in a state of siege. I do not mean that it was beleaguered by the Indians in any formal, persistent investment, but it was so constantly and so closely observed by war parties, hidden in the adjacent woods and the mountain passes, that there was little safety outside its stockade for anything less than a company of infantry or a troop of cavalry; and not always, as we shall see, for those.

Rarely in the history of the Indian wars of the United States have the Indians, no matter how preponderant in force, conducted a regular siege, Pontiac's investment of Detroit being almost unique in that particular. But they literally surrounded Fort Phil Kearney at all times. Nothing escaped their observation, and no opportunity to harass and to cut off detached parties of the garrison, to stampede the herds, or to attack the wagon trains,

* Although the general for whom this fort was named spelled his name Kearny, the name of the fort is written as above in all official documents I have examined.

was allowed to pass by. Not a stick of timber could be cut, not an acre of grass mowed, except under heavy guard. Herds of beef cattle, the horses for the cavalry and mounted infantry, the mules for the supply wagons, could not graze, even under the walls of the fort, without protection. The country teemed with game. Hunting parties were absolutely forbidden. To take a stroll outside the stockade on a summer evening was to invite death, or worse if the stroller happened to be a woman. There was no certainty about the attacks, except an assurance that one was always due at any given moment. As old James Bridger, a veteran plainsman and fur trader, a scout whose fame is scarcely less than that of Kit Carson, and the confidential companion adviser of Carrington in 1866, was wont to say to him: "Whar you don't see no Injuns thar they're sartin to be thickest."

Taking at random two average months in the two different years during which the post was maintained, one in the summer, another in the fall, I find that there were fifteen separate and distinct attacks in one and twenty in the other. In many of these, in most, in fact, one or more men were killed and a greater number wounded. Not a wagon train bound for Montana could pass up the Bozeman trail, which ran under the walls of the fort, and for the protection of which it had been established, without being attacked again and again. Only the most watchful prudence, the most skilful management, and the most determined valor, prevented the annihilation of successive parties of emigrants seeking the new and inviting land.

The war with the Indians was about the ownership of territory, as most of our Indian wars have been. Indeed, that statement is true of most of the wars of the world. The strong have ever sought to take from the weak. The westward-moving tide of civilization had at last pressed back from the Missouri and the Mississippi the Sioux and their allies, the Cheyennes, the largest and most famous of the several great groups of Indians who have disputed the advance of the white man since the days of Columbus, saving perhaps the Creeks and the Iroquois.

The vast expanse of territory west of the hundredth meridian, extending from the Red River to the British Columbia boundary line, was at the time practically devoid of white settlements, except at Denver and Salt Lake, until the Montana towns were reached in the northwest.* It is a great sweep of land which comprises every variety of climate and soil. The huge Big Horn Mountains severed that immense domain. The Sweet Water Country and all east of the Wind River Range, including South Pass and the region west of the great bend of the North Platte, had its prairies and fertile valleys. Just north of the Big Horn Mountain Range, which took in the territory which formed the most direct route to Central Montana, and the occupation of which was the real objective of Carrington's expedition in the spring of 1866, was the most precious section, controlled by tribes jealous of any intrusion by the whites.

All along the Yellowstone and its tributaries, in spite of the frequent "*Mauvaises Terres*," or "bad lands," of apparent volcanic origin, the whole country was threaded with clear streams from the Big Horn Range. The valleys of these were luxuriant in their natural products and their promise. Enormous herds of buffalo

* The country is roughly comprehended by the boundary lines between which Mountain Standard, or 105th meridian, Time, prevails.

roamed the plains, affording the Indian nearly everything required for his support. The mountains abounded with bear, deer and other game in great variety. The many rivers which traversed the territory teemed with fish, the valleys which they watered were abundantly fertile for the growing of the few crops which the Indian found necessary for his support. The land was desirable naturally and attracted the attention of the settlers.

It cannot be gainsaid that the Indians enjoyed a quasi-legal title to this land. But if a comparatively small group of nomadic and savage tribes insists upon reserving a great body of land for a mere hunting ground, using as a game preserve that which, in a civilized region, would easily support a great agricultural and urban population of industrious citizens seeking relief from the crowded and confined conditions of older communities, what are you going to do about it? Experience has shown that in spite of treaties, purchases and other peaceful means of obtaining it, there is always bound to be a contest about that land. The rights of savagery have been compelled to yield to the demands of civilization, ethics to the contrary notwithstanding. And it will always be so, sad though it may seem to many.

The close of the Civil War threw many soldiers out of employment. After four years of active campaigning they could not settle down to the humdrum life of village and country again. With a natural spirit of restlessness they gathered their families, loaded their few household belongings into wagons, and in parties of varying sizes made their way westward. Railroads began to push iron feelers across the territory. Engineers and road builders, as well as emigrants, demanded the protection of the government. At first most of the settlers merely wished to pass through the country and settle in the fair lands upon the other side, but the fertility and beauty that met their eyes on every hand irresistibly invited settlement on the journey.

At that time there were four great routes of transcontinental travel in use: southward over the famous Santa Fé trail; westward over the Kansas trail to Denver; westward on the Oregon trail through Nebraska and Salt Lake City to California and Oregon; northwestward on the Bozeman trail through Wyoming to Montana. The Union Pacific road was building along the Oregon trail, the Kansas Pacific along the Kansas trail to Denver, while the great Santa Fé system was not yet dreamed of.

The railroads being in operation for short distances, the only method of transportation was in the huge Conestoga wagon, or prairie schooner which, with its canvas top raking upward fore and aft over a capacious wagon box, looked not unlike the hull of the boat from which it took its name. These wagons were drawn by four or six mules — sometimes by oxen, known as "bull teams"— and, stores there being none, carried everything that a settler was apt to need in the new land, including the indispensable wife and children.

I am concerned in this article only with the Bozeman or Montana trail.

Early in 1866 Government Commissioners at Fort Laramie, Nebraska, were negotiating a treaty with the Sioux and Northern Cheyennes to secure the right of way for emigrants through that territory which, by the Harney-Sanborne treaty, had been conceded to them in 1865. Red Cloud, an Oglala Sioux, the foremost of the young warriors, led the objectors to the treaty, even to the point of fighting, and opposed the more conservative chiefs who deprecated war as eventually fatal to all their territorial claims. During this council, to anticipate later events, Carrington, then approaching with troops, arrived in advance, dismounted, and was introduced to the members of the council. Red Cloud, noticing his shoulder straps, hotly denounced him as the "White Eagle" who had come to steal the road before the Indian said yes or no. In full view of the mass of Indians who occupied the parade ground he sprang from the platform under the shelter of pine boughs, struck his tepees and went on the warpath. A telegram by Carrington advising suspension of his march until the council came to some agreement was negatived, and although Sunday he pushed forward nine miles beyond the fort before sunset.

One stipulation upon which the United States insisted was the establishment of military posts to guard the trail, without which it was felt the treaty would amount to nothing. The Brulé Sioux, under the lead of Spotted Tail, Standing Elk and others, favored the concession, and ever after remained faithful to the whites. The older chiefs of other Sioux bands, in spite of Red Cloud's defection and departure, remained in council for some days and, although sullen in manner and noisy in protests, finally accepted valuable gifts and indemnities and so far satisfied the Commission that they despatched special messengers to notify the District Commander that "satisfactory treaties had been made with the tribes represented at Laramie and that its route was safe." Emigrant trains were also pushed forward with their assurance that an ample force of regulars had gone up the country to ensure their safety. The sequel will appear later.

II. General Carrington's Romantic Expedition

Pursuant to the plan, Brigadier-General Henry B. Carrington, Colonel of the Eighteenth Regular Infantry, was ordered with the second battalion of his regiment, about to become the Twenty-seventh Regular Infantry, to establish, organize and take command of what was known as the Mountain District. The Mountain District at that time had but one post in it, Fort Reno, one hundred and sixty miles trom Fort Laramie. Carrington was directed to march to Fort Reno, move it forty miles westward, garrison it, and then, with the balance of his command, establish another post on the Bozeman trail, between the Big Horn Mountains and the Powder River, so as to command that valley much frequented by Indians; and, lastly, to establish two other posts, one on the Big Horn, the other on the Yellowstone, for the further protection of the trail.

General Carrington was a graduate of Yale College. He had been a teacher, an engineer and scientist, a lawyer and man of affairs, a student of military matters as well as Adjutant-General of Ohio for several years prior to the outbreak of the Civil War. At the beginning of that struggle he promptly moved one battery and several regiments of Ohio Militia into West Virginia to take part in the Battle of Phillipi before the State Volunteers could be mustered into the United States service. Without his solicitation, on May 14th, 1861, he had been appointed Colonel of the Eighteenth United States Infantry, promoted Brigadier-General November 29th, 1863, and had rendered valuable and important services during the war. He was a high-minded Christian gentleman, a soldier of large experience and proven courage, an administrator of vigor and capacity, and, as his subsequent career has shown, a man of fine literary talents.* No better choice could have been made for the expedition.

After many delays, due principally to difficulties in securing transportation, a little army of seven hundred men, accompanied by four pieces of artillery, two hundred and twenty-six wagons, and a few ambulances containing the wives and children of several of the officers, set forth from old Fort Kearney, Nebraska, on the 19th of May, 1866. About two hundred of the men were veterans, the rest raw recruits. They were armed with old-fashioned Springfield, muzzle-loading muskets, save a few who had the new Spencer breech-loading carbine, a weapon of rather short trajectory, but a great improvement on the old army musket from the rapidity of fire which it permitted. A portion of the command was mounted from the discarded horses of a cavalry regiment going east to be mustered out. They were not trained horsemen, however, and at first were rather indifferent mounted infantrymen.

Among the soldiers were artificers and mechanics of every description. The government had provided appliances needed for building forts, including tools, doors, sash, glass, nails, stoves, steel, iron, mowers, reapers, scythes, and two steam sawmills. The officers were in the main a fine body of men, most of whom had learned their soldiering in the Civil War.

It seems incredible to think that women should accompany such an expedition, but no grave anticipations

* Among his literary works he is best known for his "Washington, the Soldier," and his "Battles of the American Revolution," which is the standard work of the kind. In a personal interview he told me he read some portion of the Bible in the original Greek and Hebrew every day for years. Not many army officers can say that, and very few civilians, either.

Copyright by D. F. Barry

CROW KING — AMERICAN HORSE*

RED CLOUD — GALL

GROUP OF FAMOUS WAR CHIEFS

*Killed at Slim Buttes

of trouble with the Indians were felt by any persons in authority at that time. The Sioux and Cheyennes had consented to the opening of the road, and though they demurred to the forts, they had not absolutely refused the treaty when the government insisted upon it. The expedition was not conceived or planned for war. It was supposed to be a peaceable expedition. In fact, General Sherman, who visited Fort Kearney before the troops began to march, personally advised the ladies to accompany the expedition as very attractive in its object and wholly peaceful. Had the authorities known what was to happen, a force three times as great would scarcely have been thought adequate for the purpose. But even had there been a full knowledge of the dangers incurred, the army women would have gone with their husbands.

History records no greater instances of romantic devotion than those exhibited by the army wife. She stands peculiar among American women to-day in that particular. The army woman in a hostile country risked much more than the men. Her fate when captured was terrible beyond description — one long agony of horror and shame until death put an end to it. I have talked with army officers of large experience and have read what others have said, and the universal testimony is that no woman who was ever captured by the plains Indians west of the Missouri was spared. It was commonly agreed among the officers and men of regiments accompanied by women — and fully understood by the women as well — that in the last extremity the women were to be shot by their own friends, rather than to be allowed to fall into the hands of the savages; but no such apprehension attended this march.

The army woman's knowledge of the peril in the

usual border warfare was not an imaginary one, either. As we may read in letters and books written by army wives, it was brought home to them directly again and again. After every campaign poor, wretched women of stranded and robbed emigrant trains or devastated settlements were brought into the various camps, to whom these army women ministered with loving care, and from whom they heard frightful and sickening details that froze the blood; yet the army wife herself never faltered in her devotion, never failed in her willingness to follow wherever her husband was sent. And, save for the actual campaigning in the field, the army wife was everywhere — sometimes there, too.

In this particular expedition there were several little children, from some of whom I have gleaned details and happenings. One of these lads, while at Fort Kearney before the march, became so expert with the bow and arrow in target shooting with young Pawnee Indians near the fort, that he challenged General Sherman to shoot over the flagstaff. The youngster accomplished it by lying upon his back with feet braced against the bow, and the general squarely withdrew from the contest, declining to follow the boy's ingenious artifice.

The march was necessarily a slow one and the distance great — some six hundred miles — so that it was not until the twenty-eighth of June they reached Fort Reno. There they were menaced by the Indians for the first time and every endeavor was made to stampede their herds. The officers and men were fast becoming undeceived as to the character of their expedition. To abandon Fort Reno, or to remove it, was not practicable. Carrington ordered it re-stockaded and put in thorough repair, garrisoned it from his command, and with the balance, something over five hundred, advanced farther into the unknown land on the 9th of July. On the 13th of July, 1866, he established his camp on the banks of the Big Piney Creek, an affluent of the Powder River, about four miles from the superb Big Horn Range, with snow-capped Cloud Peak towering nine thousand feet into the heavens, close at hand. A few days later, on a little, flower-decked, grass-covered plateau, bare of trees, which fortunately happened to be just the size to contain the fort he proposed to erect, and which sloped abruptly away in every direction, forming a natural glacis, he began building the stockade.

III. The Outpost of Civilization

The plateau lay between two branches of the Piney. To the eastward of the smaller branch rose a high hill called Pilot Hill. West of it was another ridge which they named Sullivant Hills. Southwest of Sullivant Hills was a high ridge called Lodge Trail Ridge, the main branch of the Piney Creek flowing between them, so that the water supply was at the eastern or "Water Gate" of the fort. The Bozeman trail passed westward, under Pilot Hill in front of the fort, crossing the Big Piney as it neared Sullivant Hills, and then, circling around Lodge Trail Ridge for easier ascent, advanced northward, twice crossing Peno Creek and its branches, before that stream joined Goose Creek, a tributary to Tongue River, one of the chief forks of the Yellowstone. The first branch of the Peno was five miles from the fort, and the second twelve miles farther, where the garrison had to cut hay, but the branch nearer the fort was especially associated with the events of December 21st, as well as with the fight of the sixth of the same month.

The spot was delightful. Adjacent to the fort were

broad stretches of fertile, brilliantly flowered, grassy, river and mountain creek valleys. The mountains and hills were covered with pines. Game there was in plenty; water was clear and abundant. Wood, while not immediately at hand, else the place would not have been practicable of defense without tremendous labor in clearing it, was conveniently adjacent.

General Carrington marked out the walls of the fort, after a survey of the surrounding country as far as Tongue River, set up his sawmills, one of them of forty horse-power, capable of cutting logs thirty inches in diameter, established a logging camp on Piney Island, seven miles distant,with no intervening hills to surmount, which made transportation easy, and began the erection of the fort. Picket posts were established upon Pilot and Sullivant Hills, which overlooked approaches both from the east and the road to the mountains. Three times Indians attempted to dislodge these pickets, once at night; but case-shot exploding over them, and each time causing loss of men or ponies, ended similar visitations.

The most careful watchfulness was necessary at all hours of the day and night. The wood trains to fetch logs to the sawmills went out heavily guarded. There was fighting all the time. Casualties among the men were by no means rare. At first it was difficult to keep men within the limits of the camp; but stragglers who failed to return, and some who had been cut off, scalped and left for dead, but who had crawled back to die, convinced every one of the wisdom of the commanding officer's repeated orders and cautions.*

* Just when the alarms were most frequent a messenger came to the headquarters, announcing that a train en route from Fort Laramie, with special messengers from that post, was corralled by Indians, and demanded immediate help. An entire company of infantry in wagons, with a mountain howitzer and several rounds of grape-shot, was

To chronicle the constant succession of petty skirmishes would be wearisome; yet they often resulted in torture and loss of life on the part of the soldiers, although the Indians in most instances suffered the more severely. One single incident may be taken as illustrative of the life of the garrison.

One afternoon, early in October, the picket reported that the wood train was attacked to the west, and shortly after signalled the approach of a small party of soldiers from the east. Detachments were sent from the post in both directions. It proved to be not a reinforcement of troops or ammunition supplies, but two ambulances with two contract surgeons and an escort of eight men, besides Bailey, the guide, and Lieutenant Grummond, who had just been appointed to the Eighteenth Infantry, and his young bride. As they approached the main gate, accompanied by the mounted men who had been sent out to meet them, they were halted to give passage to an army wagon from the opposite direction. It was escorted by a guard from a wood train, and brought in the scalped, naked, dead body of one of their comrades, a strange welcome, indeed, to the young wife, who, upon leaving Laramie, had been assured of a beautiful ride through fertile valleys without danger, and sadder yet in its sequel two months later.

Meanwhile the work of erecting the fort was continued. It was a rectangle, six hundred by eight hundred feet, inclosed by a formidable stockade of heavy pine logs standing eight feet high, with a continuous banquette, and flaring loopholes at every fourth log. There

hastened to their relief. It proved to be a train with mail from the Laramie Commission announcing the consummation of a "satisfactory treaty of peace with all the Indians of the Northwest," and assuring the District Commander of the fact. The messenger was brought in in safety, and *peace* lasted until his message was delivered. So much was gained — that the messenger did not lose his scalp en route.

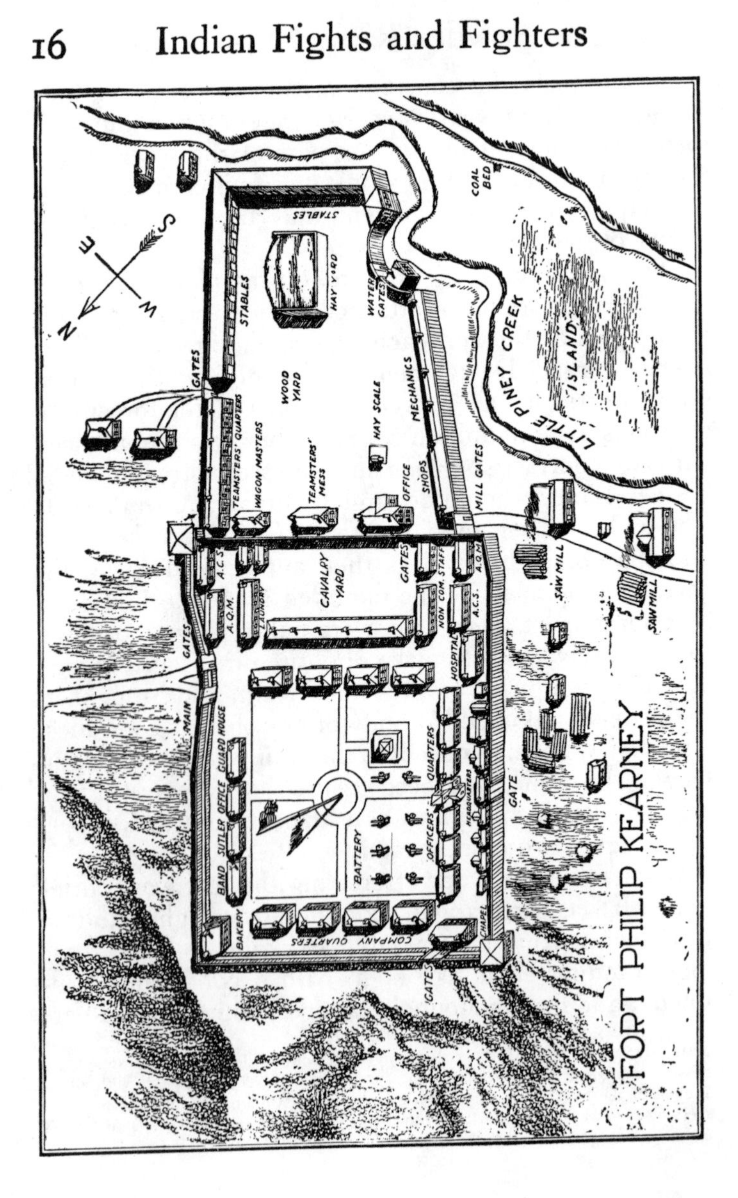
FORT PHILIP KEARNEY
LITTLE PINEY CREEK
ISLAND
COAL BED
STABLES
HAY YARD
WATER GATES
GATES
WOOD YARD
HAY SCALE
MECHANICS
TEAMSTERS' QUARTERS
WAGON MASTERS
TEAMSTERS' MESS
OFFICE
SHOPS
MILL GATES
SAW MILL
SAW MILL
A.C.S.
A.Q.M.
LAUNDRY
CAVALRY YARD
GATES
NON COM. STAFF
A.Q.M.
A.C.S.
HOSPITAL
MAIN GATES
GUARD HOUSE
SUTLER OFFICE
BAND
BAKERY
BATTERY
OFFICERS' QUARTERS
HEADQUARTERS
GATE
CHAPEL
COMPANY QUARTERS
GATES
N
E
S
W

were enfilading blockhouses on the diagonal corners, with portholes for the cannon, and quarters for officers and men, with other necessary buildings. The commanding officer's quarters was a two-story building of framed lumber, surmounted by a watch-tower. The officers' and men's quarters were built of logs. The warehouses, four in number, eighty feet by twenty-four, were framed.

East of the fort proper was a corral of slightly less area, surrounded by a rough palisade of cottonwood logs, which inclosed the wood train, hay, and miscellaneous supplies. Everything — stockade, houses, stables, in all their details, blacksmith shops, teamsters' quarters, and so on — was planned by Carrington himself.*

The main fort inclosed a handsome parade ground, in the center of which arose the tall flagstaff planned and erected by a ship carpenter in the regiment. From it, on the 31st of October, with great ceremony and much rejoicing, the first garrison flag that ever floated over the land was unfurled. The work was by no means completed as it appears on the map, but it was inclosed, and there were enough buildings ready to house the actual garrison present, although the fort was planned for a thousand men, repeatedly promised but not furnished, while all the time both cavalry and the First Battalion of the Eighteenth were held within the peaceful limits of Fort Laramie's control.

Early in August Captain Kinney, with two companions, had been sent ninety miles to the northward to establish the second post on the Big Horn, which was called Fort C. F. Smith, and was very much smaller and

* General W. B. Hazen, upon inspection of this post's stockade, pronounced it "the best he had ever seen, except one built by the Hudson Bay Company, in British America."

less important than Fort Phil Kearney. The third projected post was not established. There were not enough men to garrison the three already in the field, much less to build a fourth.

CHAPTER TWO

The Tragedy of Fort Phil Kearney

I. How the Fighting Began

TO summarize the first six months of fighting, from the first of August to the close of the year, the Indians killed one hundred and fifty-four persons, including soldiers and citizens, wounded twenty more, and captured nearly seven hundred animals—cattle, mules, and horses. There were fifty-one demonstrations in force in front of the fort, and they attacked every train that passed over the trail.

As the fort was still far from completion, the logging operations were continued until mid-winter. On every day the weather permitted, a heavily guarded train of wood-cutters was sent down to Piney Island, or to the heavier timber beyond, where a blockhouse protected the choppers. This train was frequently attacked. Eternal vigilance was the price of life. Scarcely a day passed without the lookout on Pilot Hill signalling Indians approaching, or the lookout on Sullivant Hills reporting that the wood train was corralled and attacked. On such occasions a strong detachment would be mounted and sent out to drive away the Indians and bring in the wood train — an operation which was invariably successful, although sometimes attended with loss.

Hostile demonstrations were met by prompt forays or pursuits, as circumstances permitted; and on one occasion the general pursued a band that ran off a herd nearly to Tongue River; but flashing mirrors betrayed Indian attempts to gain his rear, and a return was ordered, abandoning the stolen stock.

One expedition is characteristic of many. On the afternoon of December 6th the lookout on Sullivant Hills signalled that the wood train was attacked, and Captain (Brevet-Lieutenant-Colonel) Fetterman, the senior captain present, was detailed with a squad of forty mounted men, including fifteen cavalrymen under Lieutenants Bingham and Grummond, with Sergeant Bowers of the infantry, a veteran of the Civil War, to relieve the wood train and drive the Indians toward the Peno Valley, while Carrington himself, with about a score of mounted infantrymen, would sweep around the north side of Lodge Trail Ridge and intercept them.

The Indians gave way before Fetterman's advance, hoping to lure the troops into an ambush, but at a favorable spot they made a stand. The fighting there was so fierce that the cavalry, which by a singular circumstance was without its officers, gave way and retreated headlong across the valley toward the ridge. The mounted infantry stood its ground, and under Fetterman's intrepid leadership was making a brave fight against overwhelming odds, the number of Indians present being estimated at more than three hundred. It would have gone hard with them, however, had not Carrington and the first six of his detachment suddenly swept around a small hill or divide and taken the Indians in reverse. The general had been forced to advance under fire, and meeting the fugitive cavalry, ordered them to fall in behind his own detachment.

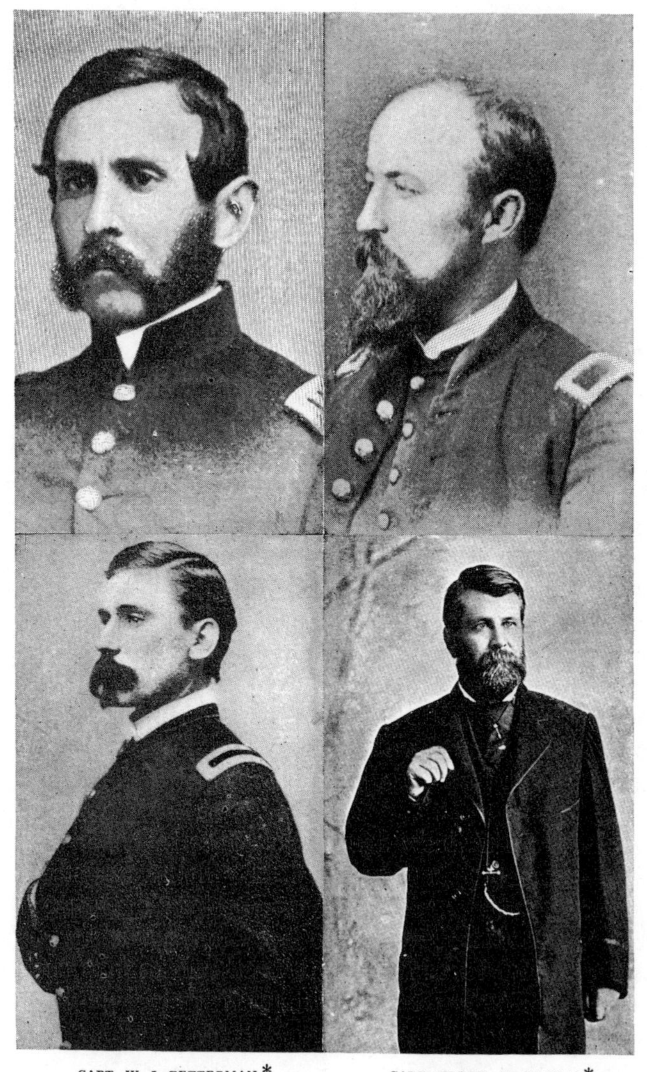

CAPT. W. J. FETTERMAN * CAPT. FREDK. H. BROWN *
LIEUT. G. W. GRUMMOND * CAPT. JAMES POWELL

THE FORT PHIL KEARNEY FIGHTERS

For portrait of General Carrington, see illustration opposite page

* Killed on Lodge Trail Ridge

He was filled with anxiety as to the course of the fight on the other side of the ridge.

Carrington, in his official report,* says: "But six men turned the point with me, one a young bugler of the Second Cavalry, who told me that Lieutenant Bingham had gone down the road around the hill to my right. This seemed impossible, as he belonged to Fetterman's command. I sounded the recall on his report, but in vain. One of my men fell and his horse on him. The principal chief operating during the day attempted to secure his scalp, but dismounting, with one man to hold the horses and reserving fire, I succeeded in saving the man and holding the position until joined by Fetterman twenty minutes later. The cavalry that had abandoned him had not followed me, though the distance was short; but the Indians, circling round and yelling, nearly one hundred in number, with one saddle emptied by a single shot fired by myself, did not venture to close in."

The rear detachment and Fetterman soon joined, and by the efforts of the combined parties the Indians were compelled to flight. It was a close call for all, but Lieutenants Grummond and Bingham were yet unaccounted for. Search was instantly made for these two officers and the infantry sergeant, who had become separated from their command while chasing some scattered Indians. One of the officers, Lieutenant Bingham, was dead. Lieutenant Grummond, after a hand-to-hand fight, was closely pressed by mounted Indians and was barely rescued. Sergeant Bowers had been fearfully wounded and scalped, although he was still alive, but died immediately.† He had killed three Indians before he had been overborne. The cavalrymen, mostly recruits, were deeply ashamed of their defection, which was partly due to the incaution of their officers in

* Published by the United States Senate in 1887.

† At the burial of Sergeant Bowers, Captain Brown, who had known him during the Civil War, pinned his Army of the Cumberland badge upon his breast, and this was found when the remains were reinterred in 1878.

leaving them to pursue a few Indians, and they were burning with a desire to retrieve their reputation, which they bravely did with their lives some two weeks later.

The casualties in the little command were two killed, five wounded. A messenger was sent to the fort for an ambulance, and the command retired in good order without further sight of the Indians. Lieutenant Bingham was not the first officer killed; for, five months before, Lieutenant Daniels, riding ahead of a small party of soldiers escorting several officers and the wife of Lieutenant Wands from Fort Laramie, had been killed in full view of the party. He had been horribly tortured with a stake before he died, and the savages put on his clothing and danced on the prairie just out of range, in front of the party, which was too small to do more than stand on the defensive. Lieutenant Grummond's wife was in the fort during the fighting on the sixth of December, and her joy at her husband's safe return can be imagined.

On the eighth of December President Andrew Johnson congratulated Congress that treaties had been made at Fort Laramie, and that *all was peace in the Northwest!*

On the 19th of December, in this peaceful territory, the wood train was again attacked in force. Carrington promptly sent out a detachment under Captain Powell with instructions to relieve the wood train, give it his support, and return with it, but not to pursue threatening Indians, for experience had shown that the Indians were constantly increasing in numbers and growing bolder with every attack. Powell efficiently performed his task. The Indians were driven off, and, although he was tempted to pursue them, he was too good a soldier to disobey orders, so he led his men back in safety to the fort.

By this time all warehouses were finished, and it was estimated that one large wood train would supply logs enough for the completion of the hospital, which alone needed attention.

Impressed by Powell's report, Carrington himself accompanied the augmented train on the 20th, built a bridge across to Piney Island to facilitate quick hauling, and returned to the fort to make ready for one more trip only. No Indians appeared in sight on that date. Already several hundred large logs had been collected for winter's firewood, besides the slabs saved at the sawmill.

It cannot be denied that there was much dissatisfaction among some of the officers at Carrington's prudent policy. They had the popular idea that one white man, especially if he were a soldier, was good for a dozen Indians; and although fifteen hundred lodges of Indians were known to be encamped on the Powder River, and there were probably between five and six thousand braves in the vicinity, they were constantly suggesting expeditions of all sorts with their scanty force. Some of them, including Fetterman and Brown, "offered with eighty men to ride through the whole Sioux Nation!" While the mettle of the Sioux Nation had not yet been fairly tried by these men, Carrington was wise enough to perceive that such folly meant inevitable destruction, and his consent was sternly refused.

The total force available at the fort, including prisoners, teamsters, citizens and employees, was about three hundred and fifty — barely enough to hold the fort, should the Indians make an attack upon it. Besides which, details were constantly needed to carry despatches, to deliver the mail, to get supplies, to succor emigrant trains, and so on. The force was woefully inadequate, and the number of officers had been depleted by detachment and other causes until there were but six left.

Ammunition was running low. There were at one time only forty rounds per man available. Repeated requests and appeals, both by letter and telegram, for reinforcements and supplies, and especially for modern and serviceable weapons, had met with little consideration. The officials in the far East hugged their treaty, and refused to believe that a state of war existed; and, if it did exist, were disposed to censure the commanding officer for provoking it. In several instances presents given in the treaty at Fort Laramie were found on the persons of visiting Indians, and one captured Indian pony was heavily loaded with original packages of those presents.

Carrington had done nothing to provoke war, but had simply carried out General Sherman's written instructions, sent him as late as August, to "avoid a general war, until the army could be reorganized and increased; but he defended himself and command stoutly when attacked. Some of the officers, therefore, covertly sneering at the caution of the commander, were burning for an opportunity to distinguish themselves on this account, and had practically determined to make or take one at the first chance. Fetterman and Brown, unfortunately, were the chief of these malcontents.

II. The Annihilation of Fetterman's Command

On the 21st of December, the ground being free from snow, the air clear and cold, the lookout on Sullivant Hills signalled about eleven o'clock in the morning that the wood train had been corralled, and was again attacked in force about a mile and a half from the fort. A relief party of forty-nine men from the Eighteenth Infantry, with twenty-seven troopers from the Second Cavalry, a detachment from which, nearly all recruits and chiefly armed with muskets as their carbines had not reached Laramie, had joined the post some months before, was at once ordered out.

The command was first given to Captain Powell, with Lieutenant Grummond in charge of the cavalry. Grummond had a wife in delicate health at the post, and he was cautioned by the officers to take care not to be led into a trap, although his experience on the 6th, when he had so narrowly escaped death, was, it would seem, the best warning he could have had. This body of men was the best armed party at the post, a few of those designated carrying the Spencer repeating carbines. Each company had been directed to keep forty rounds per man on hand for immediate use in any emergency, besides extra boxes always kept in company quarters. The men had been exercised in firing recently and some of the ammunition had been expended, although they still had an abundant supply for the purposes of the expedition. Carrington personally inspected the men before they left, and rejected those who were not amply provided.

The situation of the wood train was critical, and the party was assembled with the greatest despatch. Just as they were about to start, Captain Fetterman, who had had less experience in the country and in Indian fighting than the other officers, for he had joined the regiment some time after the fort had been built and expected assignment to command Fort C. F. Smith, begged for the command of the expedition, pleading his senior captaincy as justification for his request. Carrington reluctantly acceded to his plea, which indeed he could scarcely have refused, and placed him in charge, giving him strict and positive instructions to "relieve the wood train, drive back the Indians, *but on no account to pursue the Indians beyond Lodge Trail Ridge*," and that so soon as he had performed this duty he was to return immediately to the fort.

Captain Fetterman, as has been said, had frequently expressed his contempt for the Indians, although his fight on December 6th had slightly modified his opinions. Carrington, knowing his views, was particular and specific in his orders. So necessary did he think the caution that he repeated it to Lieutenant Grummond, who, with the cavalry, followed the infantry out of the gate, the infantry, having less preparation to make, getting away first. These orders were delivered in a loud voice and were audible to many persons — women, officers, and men in the fort. The general went so far as to hasten to the gate after the cavalry had left the fort, and from the sentry platform or banquette overlooking it, called out after them again, emphatically directing them "on no account to pursue the Indians across Lodge Trail Ridge."

The duty devolved upon Captain Fetterman was exactly that which Captain Powell had performed so satisfactorily a few days before. With Captain Fetterman went Captain Brown, with two citizens, frontiersmen and hunters, as volunteers. These two civilians, Wheatley and Fisher, were both armed with the new breech-loading rapid-fire Henry rifle, with which they were anxious to experiment on the hostiles. Wheatley left a wife and children in the fort.

Captain Frederick Brown, a veteran of the Civil War, had just been promoted, had received orders detaching

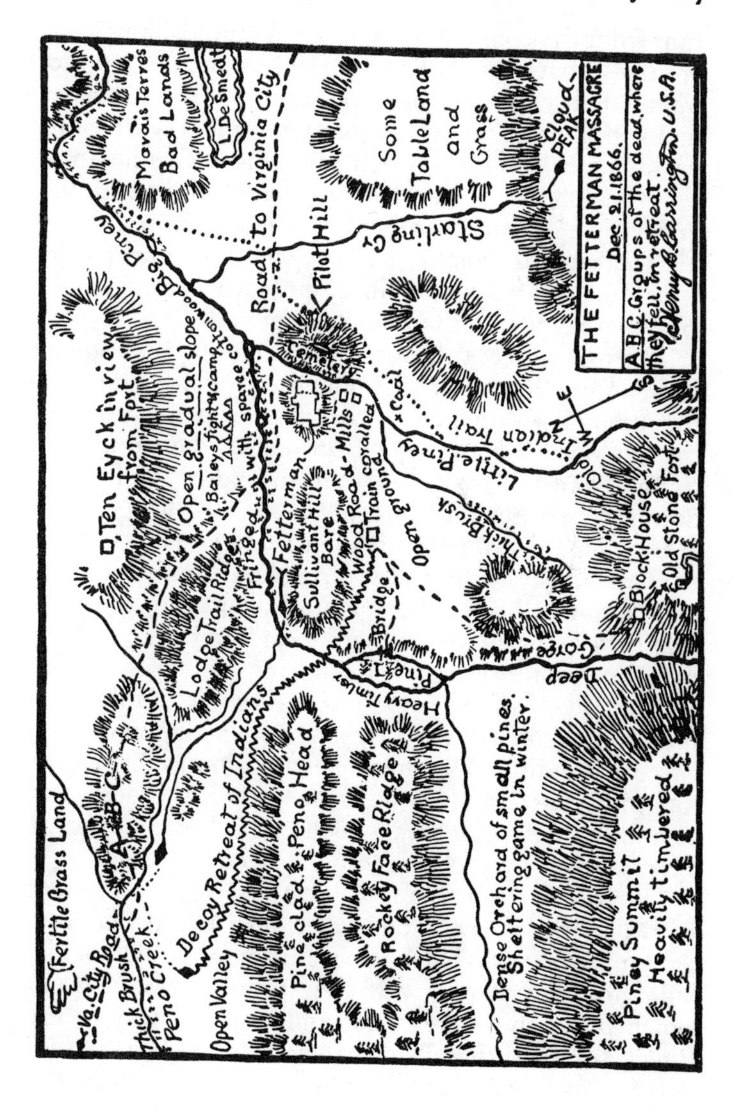
THE FETTERMAN MASSACRE
Dec. 21. 1866.
A.B.C. Groups of the dead, where they fell, in retreat.
Henry B. Carrington U.S.A.
Cloud Peak
Some Table Land and Grass
Mavais Terres Bad Lands
L. De Smedt
Road to Virginia City
Pilot Hill
Starling Cr
Big Piney
Ten Eyck in view from Fort
Open gradual slope
Cemetery
Coal
Indian Trail
Little Piney
Fetterman
Sullivant Hill Bare
Wood Road
Mills
Train coralled
Open ground
Thick Brush
Block House
Old Stone Fort
Deep Gorge
Lodge Trail Ridge
Fringed with sparse
Decoy Retreat of Indians
Pine clad
Peno Head
Rockey Face Ridge
Heavy Timber
Pine Is.
Bridge
Dense Orchard of small pines.
Sheltering game in winter.
Piney Summit
Heavily timbered
Fertile Grass Land
Va. City Road
Thick Brush
Peno Creek
Open Valley

him from the command, and was simply waiting a favorable opportunity to leave. He was a man of the most undaunted courage. His position as quartermaster had kept him on the watch for Indians all the time, and he announced on the day before the battle that he "must have one chance at the Indians before he left." It is believed, however, that his impetuous counsel, due to his good luck in many a brush with assailing parties, which he had several times pursued almost alone, largely precipitated the final disaster.

The total force, therefore, including officers and citizens, under Fetterman's command, was eighty-one—just the number with which he had agreed to ride through the whole Sioux Nation. No one in the command seems to have had the least idea that any force of Indians, however great, could overcome it.

Captain Fetterman, instead of leading his men direct to the wood train on the south side of Sullivant Hills, double-quicked toward the Peno Valley on to the north side. Perhaps he hoped that he could take the Indians in reverse and exterminate them between his own troops and the guard of the wood train—which all told comprised some ninety men—when he rounded the western end of the hills. This movement was noticed from the fort; but, as it involved no disobedience of orders, and as it might be considered a good tactical manœuver, no apprehension was felt on account of it.

The Indians surrounding the wood train were well served by their scouts, and when they found that Fetterman's force was advancing on the other side of the hill, they immediately withdrew from the wood train, which presently broke corral and made its way to the Piney, some seven miles northwest of the fort, unmolested. As Fetterman's troops disappeared down the valley, a number of Indians were observed along the Piney in front of the fort. A spherical case-shot from a howitzer in the fort exploded in their midst, and they vanished. The Indians were much afraid of the "gun that shoots twice," as they called it.

At that time it was discovered that no doctor had gone with the relieving party, so Acting-Assistant Surgeon Hines, with an escort of four men, was sent out with orders to join Fetterman. The doctor hastened away, but returned soon after with the information that the wood train had gone on, and that when he attempted to cross the valley of the Peno to join Fetterman's men he found it full of Indians, who were swarming about Lodge Trail Ridge, and that no sign of Fetterman was observed. Despite his orders, he must have gone over the ridge.

The alarm caused in the fort by this news was deepened by the sound of firing at twelve o'clock. Six shots in rapid succession were counted, and immediately after heavy firing was heard from over Lodge Trail Ridge, five miles away, which continued with such fierceness as to indicate a pitched battle. Carrington instantly despatched Captain Ten Eyck with the rest of the infantry, in all about fifty-four men, directing him to join Fetterman's command, then return with them to the fort. The men went forward on the run. A little later forty additional men were sent after Ten Eyck. Carrington at once surmised that Fetterman had disobeyed orders, either wittingly or carried away by the ardor of the pursuit, and was now heavily engaged with the Indians on the far side of the ridge.

Counting Fetterman's detachment, the guards of the wood train, and Ten Eyck's detachments, the garrison of the fort was now reduced to a very small number.

The place, with its considerable extent, might now be attacked at any time. Carrington at once released all prisoners from the guard-house, armed the quartermaster's employees, the citizens, and mustered altogether a force of only one hundred and nineteen men to defend the post.* Although every preparation for a desperate defense had been made, there were not enough men to man the walls.

The general and his remaining officers then repaired to the observatory tower, field glasses in hand, and in apprehension of what fearful catastrophe they scarcely allowed themselves to imagine. The women and children, especially those who had husbands and fathers with the first detachment, were almost crazed with terror.

Presently Sample, the general's own orderly, who had been sent with Ten Eyck, was seen galloping furiously down the opposite hill. He had the best horse in the command (one of the general's), and he covered the distance between Lodge Trail Ridge and the fort with amazing swiftness. He dashed up to headquarters with a message from Ten Eyck, stating that "the valley on the other side of the ridge is filled with Indians, who are threatening him. The firing has stopped. He sees no sign of Captain Fetterman's command. He wants a howitzer sent out to him."

* PHIL KEARNEY GARRISON,
at date of massacre, from "Post Returns":—

Wood Party, besides teamsters	55 men
Fetterman's Party (two citizens)	81 "
Ten Eyck's Party (relieving)	94 "
Helpless in hospital	7 "
Roll-call, of present, all told	119 "
Total officers and men	356 men

Ninety rifles worn out by use on horseback. Citizen employees used their private arms.

Information furnished by General H. B. Carrington.

The following note was sent to Captain Ten Eyck:

"Forty well-armed men, with three thousand rounds, ambulances, stores, etc., left before your courier came in. You must unite with Fetterman. Fire slowly, and keep men in hand. You would have saved two miles toward the scene of action if you had taken Lodge Trail Ridge. I order the wood train in, which will give fifty men to spare."

No gun could be sent him. Since all the horses were already in the field, it would have required men to haul it. No more could be spared, and not a man with him could cut a fuse or handle the piece anyway. The guns were especially needed at the fort to protect women and children.

Late in the afternoon Ten Eyck's party returned to the fort with terrible tidings of appalling disaster. In the wagons with his command were the bodies of forty-nine of Fetterman's men; the remaining thirty-two were not at that time accounted for. Ten Eyck very properly stood upon the defensive on the hill and refused to go down into the valley in spite of the insults and shouts of the Indians, who numbered upward of two thousand warriors, until they finally withdrew. After waiting a sufficient time, he marched carefully and cautiously toward Peno Valley and to the bare lower ridge over which the road ran.

There he came across evidences of a great battle. On the end of the ridge, nearest the fort, in a space about six feet square, inclosed by some huge rocks, making a sort of a rough shelter, he found the bodies of the forty-nine men whom he had brought back. After their ammunition had been spent, they had been stripped, shot full of arrows, hacked to pieces, scalped, and mutilated in a horrible manner. There were no evidences of a very severe struggle right there. Few cartridge shells lay on the ground. Of these men, only four besides the two officers had been killed by bullets. The rest had been killed by arrows, hatchets, or spears. They had evidently been tortured to death.

Brown and Fetterman were found lying side by side, each with a bullet wound in the left temple. Their heads were burned and filled with powder around the wounds. Seeing that all was lost, they had evidently stood face to face, and each had shot the other dead with his revolver. They had both sworn to die rather than be taken alive by the Indians, and in the last extremity they had carried out their vows. Lieutenant Grummond, who had so narrowly escaped on the 6th of December, was not yet accounted for, but there was little hope that he had escaped again.

III. Carrington's Stern Resolution

The night was one of wild anxiety. Nearly one-fourth of the efficient force of the fort had been wiped out. Mirror signals were flashed from the hills during the day, and fires here and there in the night indicated that the savages had not left the vicinity. The guards were doubled, every man slept with his clothing on, his weapons close at hand. In every barrack a non-commissioned officer and two men kept watch throughout the night. Carrington and the remaining officers did not sleep at all. They fully expected the fort to be attacked. The state of the women and children can be imagined, although all gossip and rumor were expressly prohibited by the commander.

The next day was bitterly cold. The sky was overcast and lowering, with indications of a tremendous

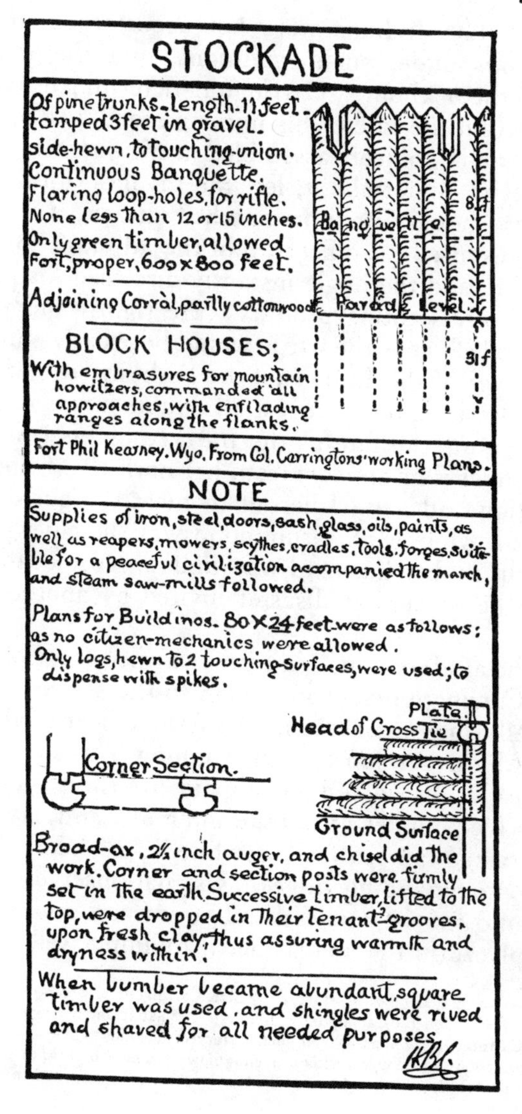
STOCKADE
Of pine trunks. length. 11 feet. tamped 3 feet in gravel.
side-hewn, to touching-union.
Continuous Banquette.
Flaring loop-holes, for rifle.
None less than 12 or 15 inches.
Only green timber, allowed
Fort, proper, 600 x 800 feet.
Adjoining Corràl, partly cottonwood.
BLOCK HOUSES;
With embrasures for mountain howitzers, commanded all approaches, with enfilading ranges along the flanks.
Banquette
Parade Level
8 f
3 f
Fort Phil Kearney. Wyo. From Col. Carrington's working Plans.
NOTE
Supplies of iron, steel, doors, sash, glass, oils, paints, as well as reapers, mowers, scythes, cradles, tools, forges, suitable for a peaceful civilization accompanied the march, and steam saw-mills followed.
Plans for Buildings. 80 X 24 feet were as follows; as no citizen-mechanics were allowed.
Only logs, hewn to 2 touching-surfaces, were used; to dispense with spikes.
Plate.
Head of Cross Tie
Corner Section.
Ground Surface
Broad-ax, 2½ inch auger, and chisel did the work. Corner and section posts were firmly set in the earth. Successive timber, lifted to the top, were dropped in their tenant-grooves, upon fresh clay, thus assuring warmth and dryness within.
When lumber became abundant, square timber was used, and shingles were rived and shaved for all needed purposes.
H.B.C.

storm. The Indians were not accustomed to active operations under such conditions, and there was no sign of them about. Carrington determined to go out to ascertain the fate of his missing men. Although all the remaining officers assembled at his quarters advised him not to undertake it, lest the savages, flushed with victory, should attempt another attack, Carrington quietly excused his officers, told the adjutant to remain with him, and the bugle instantly disclosed his purpose in spite of dissenting protests. He rightly judged that the moral effect of the battle would be greatly enhanced, in the eyes of the Indians, if the bodies were not recovered. Besides, to set at rest all doubts it was necessary to determine the fate of the balance of his command. His own wife, as appears from her narrative,* approved his action and nerved herself to meet the possible fate involved, while Mrs. Grummond was the chief protestant that, as her husband was undoubtedly dead, there should be no similar disaster invited by another expedition.

In the afternoon, with a heavily armed force of eighty men, Carrington went in person to the scene of battle. The following order was left with the officer of the day: "Fire the usual sunset gun, running a white lamp to masthead. If the Indians appear fire three guns from the twelve-pounder at minute intervals, and, later, substitute a red lantern for the white." Pickets were left on two commanding ridges, as signal observers, as the command moved forward. The women and children were placed in the magazine, a building well adapted

* "AB-SA-RA-KA, Land of Massacre," by Mrs. Carrington, of which Oliver Wendell Holmes wrote to General Carrington as follows: "What an interesting record is that of Mrs. Carrington! I cannot read such a story of devotion and endurance in the midst of privations and danger, without feeling how little most of us know of what life can be when all the human energies are called out by great enterprises and emergencies."

for defense, which had been stocked with water, crackers, etc., for an emergency, with an officer pledged not to allow the women to be taken alive, if the General did not return and the Indians overcame the stockade.

Passing the place where the greatest slaughter had occurred, the men marched cautiously along the trail. Bodies were strung along the road clear to the western end farthest from the fort. Here they found Lieutenant Grummond. There were evidences of a desperate struggle about his body. Behind a little pile of rock, making a natural fortification, were the two civilians who had been armed with the modern Henry rifle. By the side of one fifty shells were counted, and nearly as many by the side of the other brave frontiersman. Behind such cover as they could obtain nearby lay the bodies of the oldest and most experienced soldiers in Fetterman's command.

In front of them they found no less than sixty great gouts of blood on the ground and grass, and a number of dead ponies, showing where the bullets of the defenders had reached their marks,* and in every direction were signs of the fiercest kind of hand-to-hand fighting. Ghastly and mutilated remains, stripped naked, shot full of arrows—Wheatley with no less than one hundred and five in him, scalped, lay before them.

Brown rode to the death of both a little Indian "calico" pony which he had given to the general's boys when they started from Fort Leavenworth, in November, 1865, and the body of the horse was found in the low ground at the west slope of the ridge, show-

* The Indians, where possible, remove the bodies of their slain. They did during this campaign, as few dead Indians ever came into possession of the troops.

ing that the fight began there, before they could reach high ground.* At ten o'clock at night, on the return, the white lamp at masthead told its welcome story of a garrison still intact.

Fetterman had disobeyed orders. Whether deliberately or not, cannot be told. He had relieved the wood train, and instead of returning to the post, had pursued the Indians over the ridge into Peno Valley, then along the trail, and into a cunningly contrived ambush. His men had evidently fought on the road until their ammunition gave out, and then had either been ordered to retreat to the fort, or had retreated of their own motion — probably the latter. All the dead cavalry horses' heads were turned toward the fort, by the way. Fetterman and Brown, men of unquestioned courage, must have been swept along with their flying men. There may have been a little reserve on the rocks on which they hoped to rally their disorganized, panic-stricken troops, fleeing before a horde of yelling, blood-intoxicated warriors. I imagine them vainly protesting, imploring, begging their men to make a stand. I feel sure they fought until the last. But these are only surmises; what really happened, God alone knows.

The judgment of the veteran soldiers and the frontiersmen, who knew that to retreat was to be annihilated, had caused a few to hold their ground and fight until they were without ammunition; then with gun-stocks, swords, bayonets, whatever came to hand, they battled until they were cut down. Grummond had stayed with them, perhaps honorably sacrificing himself in a

* Once, while loading the bodies in the wagons, a nervous sergeant mistook one of the pickets for Indians in the rear, and gave the alarm. His detail was sharply ordered by the general to "leave their ammunition and get back to the fort as best they could, if they were afraid; for no armed man would be allowed to leave until the last body was rescued."

LIEUTENANT GRUMMOND SACRIFICING HIMSELF TO COVER THE RETREAT

Drawing by Charles Schreyvogel

vain endeavor to cover the retreat of the rest of his command. The Indian loss was very heavy, but could not exactly be determined.

IV. The Reward of a Brave Soldier

Such was the melancholy fate of Fetterman and his men. The post was isolated, the weather frightful. A courier was at once despatched to Fort Laramie, but such means of communication was necessarily slow, and it was not until Christmas morning that the world was apprised of the fatal story. In spite of the reports that had been made and fatuously believed, that peace had obtained in that land, it was now known that war was everywhere prevalent. The shock of horror with which the terrible news was received was greater even than that attendant upon the story of the disastrous battle of the Little Big Horn, ten years later. People had got used to such things then; this news came like a bolt from the blue.

Although Carrington had conducted himself in every way as a brave, prudent, skilful, capable soldier, although his services merited reward, not censure, and demanded praise, not blame, the people and the authorities required a scapegoat. He was instantly relieved from command by General Cooke, upon a private telegram from Laramie, never published, before the receipt of his own official report, and was ordered to change his regimental headquarters to the little frontier post at Fort Caspar, where two companies of his first battalion, just become the new Eighteenth, were stationed, while four companies of the same battalion, under his lieutenant-colonel, were ordered to the relief of Fort Phil Kearney.

The weather had become severe and the snow was banked to the top of the stockade. The mercury was in the bulb. Guards were changed half-hourly. Men and women dressed in furs made from wolfskins taken from the hundreds of wolves which infested the outside butcher-field at night, and which were poisoned by the men for their fur. Upon the day fixed precisely for the march, as the new arrivals needed every roof during a snow-storm which soon became a blizzard, Carrington, his wife and children, his staff and their families, including Mrs. Grummond, escorting the remains of her husband to Tennessee, and the regimental band, with its women and children, began that February "change of headquarters." They narrowly escaped freezing to death. More than one-half of the sixty-five in the party were frosted, and three amputations, with one death, were the immediate result of the foolish and cruel order.

It was not until some time after that a mixed commission of soldiers and civilians, which thoroughly investigated Carrington's conduct, having before them all his books and records from the inception of the expedition until its tragic close, acquitted him of all blame of any sort, and awarded him due praise for his successful conduct of the whole campaign. His course was also the subject of inquiry before a purely military court, all of them his juniors in rank, which also reported favorably. General Sherman expressly stated that "Colonel Carrington's report, to his personal knowledge, was fully sustained," but by some unaccountable oversight or intent, the report was suppressed and never published, thereby doing lasting injustice to a brave and faithful soldier.

At the same time the government established the sub-post between Laramie and Fort Reno, so earnestly recommended by Carrington, in October, calling it Fort Fetterman, in honor of the unfortunate officer who fell in battle on the 21st of December.

Perhaps it ill becomes us to censure the dead, but the whole unfortunate affair arose from a direct disobedience of orders on the part of Fetterman and his men. They paid the penalty for their lapse with their lives; and so far, at least, they made what atonement they could. A year later opportunity was given the soldiers at Phil Kearney to exact a dreadful revenge from Red Cloud and his Sioux for the slaughter of their brave comrades.

CHAPTER THREE

The Thirty-two Against the Three Thousand

I. The Improvised Corral on Piney Island

RED CLOUD, who had been one of the sub-chiefs of the Sioux, gained so much prestige by the defeat and slaughter of Fetterman's men that he became at once the leading war chief of the nation.* The angry braves, flushed with conquest and eager for blood, hastened to enroll themselves by thousands in his band.

Fort Phil Kearney had been in a state of siege before: it was more closely invested now than ever. Contrary to their usual avoidance of the war-path in cold weather, throughout the long and bitter winter there was no intermittence to the watchfulness of the Indians. The

* Mahapiya-luta, Red Cloud, was one of the most famous of the great Sioux Nation. He was a fierce and ruthless warrior, but withal a man of his word. After the abandonment of Fort Phil Kearney he participated in no important actions with the soldiers, although he was elected head chief of the Sioux. In the war of 1876 his camp was surprised by General Mackenzie before he had an opportunity to go on the war-path. His men were disarmed, and with him were detained in the reservation. It was a fortunate thing for the army.

Recently the old chief was asked to tell the story of his most thrilling adventure. It was a tale of one man against seven, and the old man's dim eyes grew bright and his wrinkled face lighted up with a strange light as he told it. A well-known warrior was jealous of Red Cloud, and, together with six of his followers, waylaid the young brave in a lonely spot.

Two of them were armed with rifles, the rest carried only bows and arrows, while garrison was constantly in arms. Attacks of all sorts were made with increasing frequency. The letters from the soldiers which got through to the East adequately describe their sense of the dire peril which menaced them. "*This may be my last letter*" is a frequent phrase. Travel on the trail was abandoned. As soon as possible in the spring, reinforcements were hurried up and the fort was completed, but the same state of affairs continued right along without intermission.

With the advance of summer Red Cloud gathered his warriors and determined upon a direct attack upon the fort itself. He was tired of skirmishing, stampeding stock, cutting off stragglers, etc. He wanted war, real war, and he got it! If, or when, he captured the fort, he would advance upon the other two forts in succession and so clear the country, once and for all, of the detested invaders, whose soldierly qualities he seems to have held in contempt. For the campaign he proposed he assembled no fewer than three thousand warriors, the flower of the Sioux Nation. Probably half of them were armed with firearms, Winchester rifles, Spencer repeating carbines, or old army muskets, including those that had been captured from Fetterman's party. Under cover of frequent skirmishes, which prevented much scouting on the part of the troops, Red Cloud gathered his warriors undiscovered and unmolested,

Red Cloud had a Winchester. At the first fire Red Cloud fell with a bullet in his thigh, but from where he lay he contrived to kill every one of his assailants.

The skill and courage he displayed on that occasion won for him many admiring followers, and as war after war with the whites broke out and he won fresh laurels his followers increased in numbers. He joined the various secret societies, passed through the terrible agony of the sun dance, and when, in 1866, the chiefs of the tribe signed a "peace paper," he stood out for and declared war. The fighting men flocked to his standard, and when the awful massacre in which he played so conspicuous a part occurred, he was proclaimed Chief of all the Sioux.

All the prestige he lost at Piney Island he regained upon the abandonment of the forts by the government, a most impolitic and unfortunate move.

and prepared to attack about the first of August, 1867.

The limits of the military reservation had been fixed at Washington — without adequate knowledge of the ground and in disregard of General Carrington's request and protests — so as to exclude the timber land of Piney Island, from which the post had been built, and from which the nearest and most available wood supply must be obtained. The post had been completed, but immense supplies of wood would be required for fuel during the long and severe winter. This was to be cut and delivered at the fort by a civilian outfit which had entered upon a contract with the government for the purpose. One of the stipulations of the contract was that the woodmen should be guarded and protected by the soldiers.

Wood-cutting began on the 31st of July, 1867, and Captain and Brevet-Major James Powell, commanding "C" Company, of the twenty-seventh Infantry, which was formerly a battalion of the Eighteenth a part of the command which had built the fort, and to which Fetterman and his men had belonged, was detailed with his company to guard the contractor's party. Captain Powell had enlisted in the army in 1848 as a private soldier. The Civil War had given him a commission in the regular service, and in its course he had been twice brevetted for conspicuous gallantry, once at Chickamauga and the second time during the Atlanta campaign, in which he had been desperately wounded. He had had some experience in Indian fighting before and since he came to the post, and had distinguished himself in several skirmishes, notably in the relief of the wood train, a few days before Fetterman's rashness and disobedience precipitated the awful disaster.

Arriving at Piney Island, some seven miles from the post, Powell found that the contractor had divided his men into two parties. One had its headquarters on a bare, treeless, and comparatively level plain, perhaps one thousand yards across, which was surrounded by low hills backed by mountains farther away. This was an admirable place to graze the herds of mules required to haul the wagons. As will be seen, it could also be turned into a highly defensive position. The other camp was in the thick of the pine wood, about a mile away across the creek, at the foot of the mountain. This division of labor necessitated a division of force, which was a misfortune, but which could not be avoided.

Powell sent twelve men under a noncommissioned officer to guard the camp in the wood, and detailed thirteen men with another noncommissioned officer to escort the wood trains to and from the fort. With the remaining twenty-six men and his lieutenant, John C. Jenness, he established headquarters on the plain in the open.

The wagons used by the wood-cutters were furnished by the quartermaster's department. In transporting the cordwood, the woodmen made use of the running gears only, the wagon bodies having been deposited in the clearing. In order to preserve their contents and to afford as much protection as possible to their occupants in case of Indian attacks, the quartermaster's department was in the habit of lining the wagon beds with boiler iron; and, to give their occupants an opportunity to fight from concealment, loopholes were cut in the sides. Almost every authority who has written of the fight has concluded that the particular wagon beds in question were so lined. This is a mistaken though natural conclusion. In a letter to an old comrade who wrote an account of the subsequent action,* Powell makes no mention of any iron lining, and it is certain that the wagons were not lined, but were just the ordinary wooden wagon beds.†

There were fourteen of these wagon bodies. Powell arranged them in the form of a wide oval. At the highest point of the plain, which happened to be in the center, this corral was made. The wagon beds were deep, and afforded ample concealment for any one lying in them. I sometimes wonder why Powell did not stand these beds on their sides instead of their bottoms, making a higher and stouter inclosure, the bottoms being heavier than the sides; but it is clear that he did not. There were plenty of tools, including a number of augers, in the camp, and with these Powell's men made a number of loopholes about a foot from the ground, in the outward sides of the wagons.

At the ends of the oval, where the configuration of the ground made it most vulnerable for attack, especially by mounted men, two wagons complete — that is, with bodies and running gears — were placed a short distance from the little corral. This would break the force of a charge, and the defenders could fire at the attacking party underneath the bodies and through the wheels. The spaces between the wagon bodies were filled with logs and sacks of grain, backed by everything available that would turn a bullet. The supplies for the soldiers and wood party were contained in this corral.

Instead of the old Springfield muzzle-loading musket, with which the troops mainly had been armed up to this

*General Rodenbough, in "Sabre and Bayonet."

† This statement is corroborated by private letter from a veteran soldier in the United States Army, who is one of the few survivors of the battle. Surgeon Horton, who was at the post from its establishment until it was abandoned, also says that the wagon beds were of ordinary boards, without lining or other protection.

time, Powell's men were provided with the new Allen modification of the Springfield breech-loading rifle. He

The Wagon-Box Corral on Piney Island

had enough rifles for his men and for all the civilian employees, and a large number of new Colt revolvers, with plenty of ammunition for all. The new rifle had never been used by the troops in combat with the Indians, and the latter were entirely ignorant of its tremendous range and power and the wonderful rapidity of fire which it permitted. They learned much about it in the next day or two, however. A quantity of clothing and blankets was issued to the troops at the fort on the first of August, and supplies for Powell's men were sent down to him.

II. The Wild Charge of the Sioux

Having matured his plans, Red Cloud determined to begin his attack on Fort Phil Kearney by annihilating the little detachment guarding the train.* Parties of Indians had been observed in the vicinity for several days, but no attack had been made since Powell's arrival until the second of August, when, about nine o'clock in the morning,† a party of some two hundred Indians endeavored to stampede the herd of mules. The herders, who were all armed, stood their ground and succeeded for the time being in beating back the attack. While they were hotly engaged with the dismounted force, sixty mounted Indians succeeded in getting into the herd and running it off. At the same time five hundred other Indians attacked the wood train at the other camp.

The affair was not quite a surprise, for the approach of the Indians had been detected and signalled from the corral on the island a few moments before. In the face

* On the same day an attack was made in force on Fort C. F. Smith, on the Big Horn.

† Powell's official report says nine, although a private letter written some time later makes the hour seven. It isn't material, anyway: there was ample time for all the fighting both sides cared for before the day was ended.

of so overwhelming a force the soldiers and civilians at the wood train immediately retreated, abandoning the train and the camp. Here four of the lumbermen were killed. The retreat, however, was an orderly one, and they kept back the Indians by a well-directed fire.

Meanwhile the herders, seeing the stampede of the mules, made an effort to join the party retreating from the wood train. The Indians endeavored to intercept them and cut them off. Powell, however, with a portion of his force, leaving the post in command of Lieutenant Jenness, immediately dashed across the prairie and attacked the savages in the rear. They turned at once, abandoning the pursuit of the herders, and fell upon Powell, who in his turn retreated without loss to the corral. His prompt and bold sortie had saved the herders, for they were enabled to effect a junction with the retreating train men and their guards and the soldiers and civilians, and eventually gained the fort, although not without hard fighting and some loss. One thing that helped them to get away from the Indians was that the savages stopped to pillage the camp and burn it and the train. Another thing was the presence of Powell's command, which they could not leave in the rear. After driving away the others and completing the destruction of the camp, they turned their attention to Powell's corral.

Some of the clothing that had been received the day before had not been unpacked or distributed, so it was used to strengthen the weak places in the corral. Powell's men lay down in the wagon beds before the loopholes; blankets were thrown over the tops of the beds to screen the defenders from observation and in the hope of perhaps saving them from the ill effects of the plunging arrow fire, and everything was got ready. Everybody had plenty of ammunition.

Some of the men who were not good shots were told off to do nothing but load rifles, of which there were so many that each man had two or three beside him, one man making use of no less than eight. Four civilians succeeded in joining the party in the corral — a welcome addition, indeed, bringing the total number up to thirty-two officers and men. Among this quartet was an old frontiersman who had spent most of his life hunting in the Indian country, and who had been in innumerable fights, renowned for his expertness in the use of the rifle — a dead-shot. This was the man to whom the eight guns were allotted. Powell, rifle in hand, stationed himself at one end of the corral; Jenness, similarly armed, was posted at the other, each officer watching one of the openings covered by the complete wagons, which were loaded with supplies so they could not be run off easily by hand.

While all these preparations were being rapidly made, although without confusion or alarm, the surrounding country was filling with a countless multitude of Indians. It was impossible at the time to estimate the number of them, although it was ascertained that more than three thousand warriors were present and engaged. Red Cloud himself was in command, and with him were the great chiefs of the great tribes of the Sioux, who were all represented — Unkpapas, Miniconjous, Oglalas, Brulés, and Sans Arcs, besides hundreds of Cheyennes.

So confident of success were they that, contrary to their ordinary practices, they had brought with them their women and children to assist in carrying back the plunder. These, massed out of range on the farthest hills, constituted an audience for the terrible drama about to be played in the amphitheater beneath them.

We can well imagine the thoughts of that little band of thirty-two, surrounded by a force that outnumbered them one hundred to one. Their minds must have gone back to that winter day, some seven months before, when twice their number had gone down to defeat and destruction under the attack of two-thirds of their present foemen. It is probable that not one of them ever expected to escape alive. The chances that they could successfully withstand an attack from so overwelming a number of foes of such extraordinary bravery were of the smallest. But not a man flinched, not a man faltered. They looked to their weapons, settled themselves comfortably in the wagon beds, thought of Fetterman and their comrades, and prayed that the attack might begin and begin at once. There were no heroics, no speeches made. Powell quietly remarked that they had to fight for their lives now, which was patent to all; and he directed that no man, for any reason, should open fire until he gave the order.

Some little time was spent by the Indians in making preparations, and then a force of about five hundred Indians, magnificently mounted on the best war ponies and armed with rifles, carbines, or muskets, detached themselves from the main body and started toward the little corral lying like a black dot on the open plain. They intended to ride over the soldiers and end the battle with one swift blow. Slowly at first, but gradually increasing their pace until their ponies were on a dead run, they dashed gallantly toward the corral, while the main body of the savages, at some distance in their rear, prepared to take advantage of any opening that might be made in the defenses. It was a brilliant charge, splendidly delivered.

Such was the discipline of Powell's men that not a shot was fired as the Indians, yelling and whooping madly, came rushing on. There was something terribly ominous about the absolute silence of that little fortification. The galloping men were within one hundred yards now, now fifty. At that instant Powell spoke to his men. The inclosure was sheeted with flame. Out of the smoke and fire a rain of bullets was poured upon the astonished savages. The firing was not as usual — one volley, then another, and then silence; but it was a steady, persistent, continued stream, which mowed them down in scores. The advance was thrown into confusion, checked but not halted, its impetus being too great; and then the force divided and swept around the corral, looking for a weak spot for a possible entrance. At the same moment a furious fire was poured into it by the warriors, whose position on their horses' backs gave them sufficient elevation to enable them to fire over the wagon beds upon the garrison. Then they circled about the corral in a mad gallop, seeking some undefended point upon which to concentrate and break through, but in vain. The little inclosure was literally ringed in fire. Nothing could stand against it. So close were they that one bullet sometimes pierced two Indians.*

Having lost terribly, and having failed to make any

* "I know that my husband never expected to come out of that fight alive. He has told me that during the fight the Indians came up so close to the corral, that one shot would pass through the Indian in advance and kill or wound the one behind. My husband claimed the honor of killing Red Cloud's nephew." — Letter from Mrs. Annie Powell to me. Surgeon Horton states that the men told him on their return to the fort that the Indians were crowded so closely together that the conical bullets from their muskets killed four or five Indians in line behind one another. The Indians came up in solid masses on every side.

CHARGE OF RED CLOUD ON THE CORRAL AT PINEY ISLAND

Drawing by R. Farrington Elwell

impression whatever, the Indians broke and gave way. They rushed pell-mell from the spot in frantic confusion till they got out of range of the deadly storm that swept the plain. All around the corral lay dead and dying Indians, mingled with killed and wounded horses kicking and screaming with pain, the Indians stoically enduring all their sufferings and making no outcry. In front of the corral, where the first force of the charge had been spent, horses and men were stretched out as if they had been cut down by a gigantic mowing-machine. The defenders of the corral had suffered in their turn. Lieutenant Jenness, brave and earnest in defense, had exposed himself to give a necessary command and had received a bullet in his brain. One of the private soldiers had been killed and two severely wounded. The thirty-two had been reduced to twenty-eight. At that rate, since there were so few to suffer, the end appeared inevitable. The spirit of the little band, however, remained undaunted. Fortunately for them, the Indians had met with so terrible a repulse that all they thought of for the time being was to get out of range. The vicinity of the corral was thus at once abandoned.

III. Red Cloud's Baptism of Fire

Red Cloud determined, after consultation with the other chiefs, upon another plan which gave greater promise of success. Seven hundred Indians, armed with rifles or muskets and followed by a number carrying bows and arrows, were told off to prepare themselves as a skirmishing party. Their preparations were simple, and consisted of denuding themselves of every vestige of clothing, including their war shirts and war bonnets. These men were directed to creep forward, taking advantage of every depression, ravine, or other cover, until they were within range of the corral, which they were to overwhelm by gun and arrow fire. Supporting them, and intended to constitute the main attack, were the whole remaining body of the Indians, numbering upward of two thousand warriors.

With the wonderful skill of which they were masters, the skirmishing party approached near to the corral and began to fire upon it. Here and there, when a savage incautiously exposed himself, he was shot by one of the defenders; but in the main the people of the corral kept silent under this terrible fusillade of bullets and arrows. The tops of the wagon sides were literally torn to pieces; the heavy blankets were filled with arrows which, falling from a distance, did no damage. The fire of the Indians was rapid and continuous. The bullets crashed into the wood just over the heads of the prostrate men, sounding like cracking thunder; yet not one man in the wagon beds was hurt. Arguing, perhaps from the silence in the corral, that the defenders had been overwhelmed and that the time for the grand attack had arrived, signal was given for the main body of the Indians to charge.

They were led by the nephew of Red Cloud, a superb young chieftain, who was ambitious of succeeding in due course to the leadership now held by his uncle. Chanting their fierce war songs, they came, on arranged in a great semicircle. Splendid, stalwart braves, the flower of the nation, they were magnificently arrayed in all the varied and highly-colored fighting panoply of the Sioux. Great war bonnets streamed from the heads of the chiefs, many of whom wore gorgeous war shirts; the painted bodies of others made dashes of rich color against the green grass of the clearing and the dark pines of the hills and mountains behind. Most of them carried on their left arms painted targets or shields of buffalo hide, stout enough to turn a musket shot unless fairly hit.

Under a fire of redoubled intensity from their skirmishers they broke into a charge. Again they advanced in the face of a terrible silence. Again at the appointed moment the order rang out. Again the fearful discharge swept them away in scores. Powell's own rifle brought down the dauntless young chief in the lead. Others sprang to the fore when he fell and gallantly led on their men. Undaunted, they came on and on, in spite of a slaughter such as no living Indian had experienced or heard of. The Indians could account for the continuous fire only by supposing that the corral contained a greater number of defenders than its area would indicate it capable of receiving. So, in the hope that the infernal fire would slacken, they pressed home the attack until they were almost at the wagon beds. Back on the hills Red Cloud and the veteran chiefs, with the women and children, watched the progress of the battle with eager intensity and marked with painful apprehension the slaughter of their bold warriors.

The situation was terribly critical. If they came on a few feet farther the rifles would be useless, and the little party of twenty-eight would have to fight hand-to-hand without reloading. In that event the end would be certain; but just before the Indians reached the corral, they broke and gave way. So close had they come that some of the troopers in their excitement actually rose to their knees and threw the augers with which the loopholes had been made, and other missiles, in the faces of the Indians. Others, however, kept up the fire, which was indeed more than mortal humanity could stand.

What relief filled the minds of the defenders, when they saw the great force which had come on so gallantly reeling back over the plains in frantic desire to get to cover, can easily be imagined. Yet such was the courage, the desperation of these Indians, in spite of repulse after repulse and a slaughter awful to contemplate, that they made no less than six several and distinct charges in three hours upon that devoted band. After the first attack made by the men on horseback, not a single casualty occurred among the defenders of the corral. It was afternoon before the Sioux got enough.

The Indians could not account for this sustained and frightful fire which came from the little fort, except by attributing it to magic. "The white man must have made bad medicine," they said afterwards, before they learned the secret of the long-range, breech-loading firearm, "to make the guns fire themselves without stopping." Indeed, such had been the rapidity of the fire that many of the gun-barrels became so hot that they were rendered useless. To this day the Indians refer to that battle as "the bad medicine fight of the white man."

The ground around the corral was ringed with Indian slain. They were piled up in heaps closer by, and scattered all over the grass farther away. Nothing is more disgraceful in the eyes of an Indian chieftain or his men than to permit the dead bodies of those killed in action to fall into the hands of the enemy. Red Cloud recognizing the complete frustration of his hopes of overwhelming Fort Phil Kearney and sweeping the invaders out of the land at that time, now only wished to get his dead away and retreat. In order to do so he threw forward his skirmishers again, who once more poured a heavy fire on the corral.

This seemed to Powell and his exhausted men the precursor of a final attack, which they feared would be the end of them. Indeed Powell, in his report, says that another attack would have been successful. From the heat and the frightful strain of the long period of steady fighting, the men were in a critical condition. The ammunition, inexhaustible as it had seemed, was running low; many of the rifles were useless. They still preserved, however, their calm, unbroken front to the foe, and made a slow, deliberate, careful reply to the firing that was poured upon them.

Red Cloud, however, had no thought of again attacking. He only wanted to get away. Under cover of his skirmishers he succeeded in carrying off most of the dead, the wounded who were able to crawl getting away themselves. A warrior, protecting himself as well as he could with the stout buffalo-hide shield he carried, would creep forward, attach the end of a long lariat to the foot of a dead man, and then rapidly retreating he would pull the body away. All the while the hills and mountains resounded with the death chants of the old men and women.

At the close of these operations a shell burst in the midst of the Indian skirmishers, and through the trees off to the left the weary defenders saw the blue uniforms of approaching soldiers, who a moment afterwards debouched in the open.

An astonishing sight met the eyes of the relief party. Clouds of Indians covered the plain. The little corral was still spitting fire and smoke into the encircling mass. They had got there in time then. Without hesitation the troops deployed and came forward on the run. Their cheers were met by welcoming shouts from Powell and his heroic comrades.

The herders, woodsmen, and guards who had escaped from their camp in the morning, had reached the fort at last with the news of Powell's imminent danger. Major Smith, with one hundred men and a howitzer, was at once despatched to his support. No one dreamed that the force of Indians was so great, or perhaps more men would have been sent, although the number at the fort was still insufficient to permit of the detachment of a very large party. It was now three o'clock in the afternoon. The Indians, disheartened and dismayed by their fearful repulse, sullenly retreated before the advance of the charging soldiers. There was a splendid opportunity presented to them to wipe out Smith's command with their overwhelming force, for they could have attacked him in the open; but they had had enough for that day, and the opportunity was not embraced.

Major Smith realized instantly that the proper thing for him to do, in the face of such great odds, was to get Powell's men and return with all speed. Carrying the bodies of the dead and wounded, the little band of defenders joined the rescuers and returned to the fort, leaving the barren honors of the field to the Indians, who occupied it on the heels of the retiring soldiers.*

IV. After the Battle. The Scout's Story

Powell modestly estimated he had killed sixty-seven Indians and wounded one hundred and twenty. Most of his men declared the Indian loss to have been between three and four hundred, but it was not until a

* Dr. Horton writes me that when Powell's men reached the post they were literally crazed with excitement and the nervous strain of the fight. The health of many of them was completely broken. Powell himself never fully recovered from the strain of that awful day, his wife informs me.

year after the battle that the real facts were ascertained from the Indians themselves. The loss in killed and wounded in the engagement, on the part of the Indians, was *one thousand one hundred and thirty-seven.* In other words, each of the defenders had accounted for at least thirty-six of the Indians. Amply, indeed, had the little band avenged the death of their comrades under Fetterman.

As Colonel Dodge justly says, the account reads like a story of Cortes.* At first sight it appears to be incredible. In explanation of it, the following account, which Colonel Dodge has preserved of a subsequent conversation between the frontiersman to whom the eight guns were allotted and the department commander is of deep interest:—

"How many Indians were in the attack?" asked the General.

"Wall, Gin'r'll, I can't say fer sartin, but I think thar wur nigh onto three thousand uv 'em."

"How many were killed and wounded?"

"Wall, Gin'r'll, I can't say fer sartin, but I think thar wur nigh onto a thousand uv 'em hit."

"How many did you kill?"

"Wall, Gin'r'll, I can't say; but gi'me a dead rest, I kin hit a dollar at fifty yards every time, and I fired with a dead rest at more'n fifty of them varmints inside of fifty yards."

"For Heaven's sake, how many times did you fire?" exclaimed the astonished General.

"Wall, Gin'r'll, I can't say, but I kept eight guns pretty well het up for more'n three hours."

On this occasion Powell received his third brevet for heroism and distinguished conduct on the field.

The next fall a new treaty was made with the Indians, and the post which had been the scene alike of heart-

* "Our Wild Indians," by Colonel R. I. Dodge, U. S. A. Mrs. Powell, in a letter to me, also vouches for the anecdote quoted.

breaking disaster and defeat and of triumph unprecedented, was abandoned to them. The troops were withdrawn. The Indians at once burned it to the ground. It was never reoccupied, and to-day is remembered simply because of its association with the first and, with one exception, the most notable of our Indian defeats in the west, and with the most remarkable and overwhelming victory that was ever won by soldiers over their gallant red foemen on the same ground.

At this writing (September, 1903) the once mighty Red Cloud, now in his eighty-ninth year, is nearing his end, and already various claimants for the now practically empty honor of the Head-Chieftainship of the Sioux have arisen, the two most prominent candidates being young Red Cloud and the son of old Sitting Bull.

Note

Since the first publication of this article I have received the following letter, which, as it tends to confirm what seems incredible, the terrible Indian loss, I quote in full:

Dear Mr. Brady:

Although I am much nearer three score than fifty, I still enjoy historical romance and facts, and I have, I think, read most of your writings. I have just read your last article and it recalls a conversation with Red Cloud twenty years ago.

He was with my dear old friend, "Adirondack Murray" and, I think, J. Amory Knox and myself. He, Murray and Knox had been photographed in a group. In reminiscing in regard to the Piney Island battle, he said he went in with over three thousand braves and lost over half. Murray asked him if he meant over fifteen hundred had been killed then, and he said:

"I lost them. They never fought again."

He knew Murray, Knox and myself wielded the pen sometimes but that we never used private talks. I tell you the above for your personal satisfaction. Sincerely, W. R. E. Collins,

3-22-'04. 1438 Broad—Exchange, New York.

CHAPTER FOUR

Personal Reminiscences of Fort Phil Kearney and the Wagon-Box Fight

By Mr. R. J. Smyth.*

"Cherokee, Ia., 6–27–1904.

AS I was a member of the Carrington Powder River Expedition of 1866, I take the liberty of sending you a short sketch of happenings about Fort Phil Kearney. Being actively engaged with others for some two years in making the history of that place, I think that the account may be of interest.

I left Fort Leavenworth early in the spring of 1866. At Fort Kearney, Nebraska, we found Col. Carrington and a part of his command, consisting of several companies of the Eighteenth Regular Infantry. Early in April we received some recruits for said command, and in a short time started on what at that time was called the Carrington Powder River Expedition. We followed the overland trail (sometimes called the Salt Lake trail) up the south side of the South Platte River to Julesberg, crossed the river there, then crossed the

* The serial publication of these articles brought me many letters filled with corrections, suggestions, and other material, written by participants in the events described. Among them all none is more graphic and more interesting than this from Mr. Smyth, formerly Teamster with Carrington, which I count it a privilege to insert in this book in his own words. — C. T. B.

divide to the North Platte. From here we went to Fort Laramie. From this point we marched west to Mussa ranch, crossed Horse Creek, and followed the Bozeman trail. This was a new road, and a short cut to Montana. After following this trail fifteen miles we struck the North Platte at Bridger's Ferry. We crossed here in a ferryboat — a large flat boat attached to a large cable rope stretched across the river.

We followed the North Platte River up on the right side to a point opposite to the present site of Fort Fetterman. At this point we left the river and struck across the country, crossing Sand Creek and several other small creeks, among which I now remember the North, South, and Middle Cheyennes. They were then merely the dry beds of what would be quite large rivers at the time of the melting of the snow in the mountains. At a point twenty-two miles east of the Powder River we struck the head of the Dry Fork of the Powder River and followed it down to the river.

There on the west side we found Fort Reno, established by General Conner in '65 and garrisoned by a few "galvanized soldiers." The garrison had been greatly reduced by desertions during the winter, the soldiers making for Montana. "Galvanized soldiers" was a name given to captured Rebel soldiers who enlisted in the Union Army to do frontier duty in order to get out of prison, and incidentally to draw pay from Uncle Sam. We laid over here for a few days, and on the fourth of July the Indians stampeded the stock of Al. Leighton, the sutler. The colonel made a detail of soldiers and citizens to go out after the Indians and recover the stock if possible.

It was indeed a laughable sight to see the soldiers trying to ride mules that were not broken to ride — and

Courtesy of The Century Co.

"BOOTS AND SADDLES:" A START IN THE EARLY MORNING

Drawing by Frederic Remington

the soldiers knew about as much about riding as the mules did. We followed the Indians to the Pumpkin Buttes and I am free to say for myself that I was very glad that we did not find them. Had we got in touch with them we would have had the smallest kind of a show to save our hair. The soldiers being mounted on green mules, and being armed with the old Springfield musket, and that strapped on their backs, a very few Indians could have stampeded the mules and, in fact, the soldiers as well.

We, the citizens, had made arrangements that if the Indians attacked us we would stick together and fight it out the best that we could. Jim Bridger, our guide, was with this party. He was an old timer in the mountains. I had two years experience in the mountains and plains prior to this time; the rest of the citizens were good men. We returned to the fort safely but did not recover any of the stock.

A day or two later we left the fort. The first day's march was a very hard one, thirty-six miles to Crazy Woman's Fork. This creek was a very fine one, clear, cool, and very rapid. The command was badly demoralized by this long, hot, and dry march, no water between that point and Fort Reno. The soldiers had been paid off a day or two before, many had been drunk, many more thoughtless, and did not provide for water in spite of orders. I saw five dollars paid for a canteen of water on this march. On our arrival at Crazy Woman's Fort, the commanding officer detailed a guard to keep the soldiers from jumping into the creek and drinking too much water.

We laid over here two days, to repair wagons and bring in the stragglers. Had the Indians been on hand, they could have cleaned up many of the soldiers at this time. From this creek west to the Big Horn the country is very fine; plenty of wood, water, and grass; in fact, a paradise. We traveled west to the forks of the Pineys. The big and little Pineys fork near where we made our camp, sixty-five miles west of the Powder River.

On the twenty-fourth day of July we moved to the place where we established Fort Phil Kearney. The grasshoppers were so thick in the air that day that they nearly obscured the sun from sight.* In fact, it did not look bigger than a silver dollar. The fort was built about as you have described it, and from the day that we established it until I left there, in November, '67, the Indians were very much in evidence and plenty of fighting nearly all that time.

I was a teamster on this expedition, driving an ambulance team. Made several trips to Fort Laramie and to Fort C. F. Smith on the Big Horn. This latter Fort was established by Carrington a short time after the establishment of Fort Phil Kearney, and was a two-company post.

I was with the hay-making party down the Big Piney during a part of the summer of '66. During one of our trips to the hay field, we were accompanied by a man who represented *Frank Leslie's Illustrated Weekly* as an artist. This man rode with me a part of the way. He intended to do some sketching near there but I advised him to stay with our outfit. However, he insisted on stopping by the way. On our return we found him dead, a cross cut on his breast, which indicated that they thought him a coward who would not fight. He wore long, black hair and his head had been completely skinned. Probably it was the work of a band of young Cheyenne bucks; they could cut the scalps into many

* I have observed similar visitations in other parts of the West years ago.—C. T. B.

pieces and thereby make a big show in camp. Was very sorry for this man; he appeared to be a perfect gentleman. His thought was, that if the Indians found him they would not hurt him, as he intended to show them his drawings, and also explain to them that he was not armed.

Later on the Indians got so thick that we had to

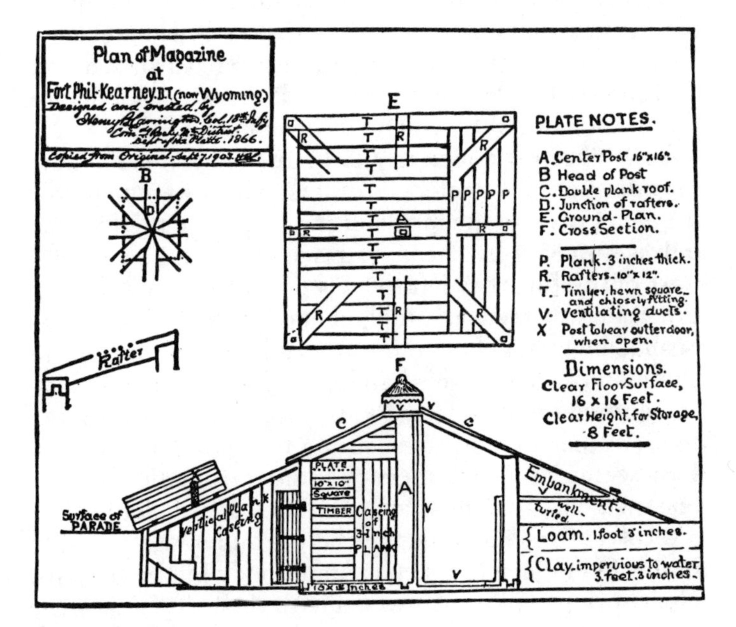

abandon this hay-making business. The day that we broke camp we had a great deal of fighting with the Indians. I remember a soldier named Pate Smith who borrowed a revolver from me that day. This man was mounted. He rode too far ahead of the outfit, the Indians cut him off. Later we heard from the Crows that the Sioux caught him, skinned him alive. This man was an old volunteer soldier, but what show has a man with the old-fashioned Springfield musket? One shot and you are done.

I was at the Fort at the time of the Phil Kearney massacre and went down with the reinforcements to that sad scene. Our men were all down when we got there, and cut up in the most brutal manner, such as only a red brute would do. We buried them a little east of the fort. They fought a good fight, but were surprised and overpowered. As we approached the scene of action the country was black with Indians to the west.

The officers were clearly to blame for this slaughter; they disobeyed the colonel's orders, which were to guard the wood train to the fort, and not to engage the Indians unless attacked by them. At a point about two miles west of the fort they left the wood train; crossed the Big Piney Creek; got nearly to the Peno Creek, and were ambushed by about three thousand Indians, and the entire command killed. This band of Indians included all of the different tribes of the Sioux, also Cheyennes, Blackfeet, Arapahoes, and some young renegade Crow bucks. I knew this latter statement to be true, from the fact that one member of Company C, Second Cavalry, had stolen a revolver from me some time before and it was with him in this fight. It was taken from his body by the Indians. Next spring a young Crow came to the fort. I saw the gun under his blanket and took it away from him. If he was in camp on the Big Horn with his people, he could not have got this gun on this field of slaughter. I had been wounded about six weeks prior to this fight and had not reported for duty, but on call for volunteers to reinforce the Fetterman party, reported for duty and went with the command to the scene of the massacre.

You are in error in stating that there was no communication with the outside world during this winter. I made one trip with my ambulance to Fort Laramie. We had an escort of ten cavalry soldiers. We made, I think, three trips after this without an escort, using pack mules, the party consisting of two packers and the mail-carrier, Van Volsey, a very fine man and a brave one, too. Last trip up I saw Indian signs in the dry forks of the Powder River, consisting of the remains of a camp fire, not entirely burned out, and some Indian traps lying around it. I refused to make another trip without an escort. On our arrival at the fort we reported the facts, and demanded an escort for the next trip. But owing to the fact that the stock was in such poor condition on account of the scarcity of food, they could not furnish one mounted.

They persuaded me to take one more trip with Van Volsey, which I foolishly consented to do. On the first day out I got snow blind and on our arrival at Fort Reno requested him to get a substitute in my place. He refused to do so and insisted that I accompany him to Fort Laramie, but after being on the road a short time, my eyes played out entirely. I had to return to the fort and there secured another man to take my place. They made the trip down all right and returning were accompanied by two or three soldiers, who were going to join their commands. They had got nearly to the head of the dry fork of the Powder River when the Indians killed the entire party. We found the bones of the men and mules and some of the mail sacks. We buried the men's remains there.

During the summer of '67 life was one continual round of fighting. We lost a great many men, but damaged the Sioux much more than during the previous year. The soldiers had better guns, and were far better Indian fighters. They had learned that it was safer to keep their faces to the Indians, than, as during the previous year, their backs. When you run from an Indian you are his meat.

On the day of the wagon-box fight, accompanied by my partner, I left the fort before daylight. We went to the foot-hills to get some deer. A short time after daylight we discovered a lot of Indian smoke signals on the hills, and decided that we had better get back to the fort. In making our way back we followed the Little Piney down for some distance, and found that the country was full of Indians. We then struck out for the wood train. The Indians had got between us and it. We then went to the wagon-box corral, and got there none too soon.

Your description of the corral is correct as I remember it to be. Its location is about right, except that it was not on an island. I never heard of Little Piney Island, and I do not believe such an island existed there at that time.* The wagon boxes were of the ordinary government boxes. They were set off from the wagons, as the wagons were in corral. The intervals between were packed with logs, bales of blankets, clothing, sacks of corn, etc. As to the wagon boxes being lined with iron, you are right. They were not. Up to that time, and during my time on the plains, I never saw wagon boxes so lined. The wagon box that I was detailed to fight in had no such protection, but we had gunny sacks of corn placed on edge two deep on the inside of the box, with a two inch auger hole at the point where the four sacks came together. This made good protection for

* General Carrington's map on page 27 shows the island. Mr. Smyth's recollection is in error here.—C. T. B.

the body when lying down. As stated in your article, the tops of the wagon boxes were literally torn to pieces with the bullets fired at us by the Indians. Without this protection the fight would not have lasted very long.

There was a surplus of ammunition and guns. I had two Spencer carbines, and two revolvers (six-shot army Colt's). During the first charge I emptied the carbines and the revolvers less two shots (reserved for myself in case of a show down). The balance of our men must have fired as many shots as I did. The soldier that was in the box with me had a needle gun and a Spencer; also one or two revolvers. And he kept them busy while he lived. This man was an infantry soldier — do not remember his company. He was shot through the head, dying in about two hours after being shot.

Lieutenant Jenness had just cautioned me not to expose my person, and to hold my fire until I was sure of getting an Indian at each shot. He had moved a few feet from my box when he was shot through the head. I think he died instantly. He was a grand, good man, and a fearless officer. I told him to keep under cover. He stated he was compelled to expose himself in order to look after his men.

I got a slight wound in my left hand; a bullet came in through my porthole, which I thought was close shooting for a Sioux.

This fight lasted about four hours, and was very hot from the start. I had been in several Indian fights prior to this time, but never saw the Indians make such a determined effort to clean us up before. They should have killed the entire party. They certainly had force enough to ride over us, but our fire was so steady and severe that they could not stand the punishment.

Our men stood the strain well, held their fire until the bullets would count. In fact, shooting into such a mass of Indians as charged on us the first time, it would be nearly impossible for many bullets to go astray. In all my experience in fighting Indians prior to this time, I never saw them stand punishment so well as they did at this time; they certainly brought all their sand with them. In charging on our little corral they rode up very close to the wagon boxes, and here is where they failed. Had they pushed home on the first charge, the fight would not have lasted ten minutes after they got over the corral.

Many dead and wounded Indians lay within a few feet of the wagon boxes. The wounded Indians did not live long after the charge was over. They would watch and try to get a bullet in on some of our men. We had to kill them for self-protection. Anyway, it was evening up the Fetterman deal. They never showed mercy to a wounded white man, and should not expect any different treatment. I had a canteen of water when the fight commenced, and used most of it to cool my guns.

You state that all of our loss occurred at time of the first charge. This is an error, as the man in my box was shot after he had been fighting nearly an hour. I think that his name was Boyle. Up to the time that he was shot he certainly filled the bill and did his duty, dying with his face to the foe as a soldier should.

I do not try to estimate the number of the Indians, but, as my partner said, "The woods were full of them." This was the largest gathering of Indians that I ever saw, and the hardest fighting lot that I ever encountered.

When the reinforcements came in sight we took on a new lease of life, and when they dropped a shell over the

Indians we knew that the fight was won. Indians will not stand artillery fire. They call it the "wagon gun." The reinforcements came just in time. One hour more of such fighting would have exhausted our men and ammunition.

As to the Indians carrying off all their dead and wounded, here you are again mistaken, as many of our men carried away with them scalps, etc., taken from the bodies of the dead Indians near the corral.* The Indians certainly hauled off all their dead and wounded that they could, but did not expose themselves very much in order to get the dead ones near the corral.

On arrival of reinforcements we immediately retreated to the fort. Captain Powell was the right man to command under such trying circumstances. No better or braver man ever held a lieutenant's commission than Jenness. As to the Indian loss, I think you have overestimated it. We thought that we had killed and wounded some more than four hundred. However, you may be right in your estimates. We had the opportunity to clean up that number, and we certainly did our best to do so.

After the massacre of '66 (Dec.) we received reinforcements, as I now remember, four companies of infantry and two companies, L and M, of the 2d Cav. This large additional force, stationed at a four-company fort, and only provisioned for four companies, caused a great deal of suffering during the winter, resulting in

* Surgeon Horton writes me that the "soldiers brought back to the fort the head of an Indian for a scientific study of Indian skulls!" He afterwards sent it to Washington. He also states that there were a number of dead bodies too near the corral for the Indians to get them during the action. When he and other officers visited the place the next day, after the withdrawal of the Indians, there were no dead bodies to be found, not even the headless one.—C. T. B.

much sickness and many deaths from scurvy. Nearly all of us were suffering from this disease. I have never fully recovered from the effect of it.

Colonel Carrington was severely censured by the War Department and many others for the Fort Phil Kearney massacre, and, I think, unjustly. Had Col. Fetterman and Capt. Brown and the other officers in command obeyed his orders, the massacre would not have occurred, not, at least, at this time.

Fetterman and Brown were dare-devil fighters, always anxious for a fight, and took this opportunity to get into one. Capt. Brown, on his "calico" pony, was a familiar figure around this fort — the boys called him "Baldy." The Indians were very anxious to kill Brown; he was a thorn in their sides. While we to some extent lay the blame of the massacre on Brown and Fetterman, to be honest, we were nearly all partly to blame. We were always harping at the colonel to send a large force out and fight the Indians, but he always insisted on a conservative course. We all thought up to that time that one hundred good men could walk through the entire Sioux Nation. This massacre demonstrated that in a fight in the open the Sioux should not have over five to one of us.

I was well posted in regard to the Carrington Powder River Expedition of 1866 and the history of Fort Phil Kearney from the date of establishment to Nov., 1867, and acquainted with all of the officers and many of the soldiers and citizens. I probably would not have written this little statement of actual history were it not for the fact that in your article you stated that you got some of the record from the only living member of the celebrated wagon-box fight. I am still in the flesh and will pull down the scale at two hundred pounds. In all probability there are others alive, as we all were young men at that time.

The history of the three forts established in 1865 and 1866, well written, would make interesting history, as almost every day was full of stirring events. Quite a number of the citizens in that country at this time were discharged volunteer soldiers and some rebel soldiers also. As a rule, they were hard nuts for the Indians to crack. It was noticeable that they would not take chances fighting citizens that they would take with the soldiers.

After leaving Fort Phil Kearney I went to Cheyenne and followed the Union Pacific R. R. to the finish. Was at Promontory Point in Utah when the Union Pacific and Central Pacific R. Rs. connected; this was one of the mile-stones in the history of the West, and practically solved the Indian problem. The Indians fought hard for this territory. It was the best hunting ground that they had left. There were many half-breeds among them, and they were daring and shifty fighters.

Respectfully yours,

R. J. SMYTH"

CHAPTER FIVE

Forsyth and the Rough Riders of '68

I. The Original "Rough Riders."

NO one will question the sweeping assertion that the grittiest band of American fighters that history tells us of was that which defended the Alamo. They surpassed by one Leonidas and his Spartans; for the Greeks had a messenger of defeat, the men of the Alamo had none. But close on the heels of the gallant Travis and his dauntless comrades came "Sandy" Forsyth's original "Rough Riders," who immortalized themselves by their terrific fight on Beecher's Island on the Arickaree Fork of the Republican River, in Eastern Colorado, in the fall of 1868.

The contagion of the successful Indian attacks on Fort Phil Kearney had spread all over the Central West. The Kansas Pacific was then building to Denver, and its advance was furiously resisted by the Indians. As early as 1866, at a council held at Fort Ellsworth, Roman Nose, head chief of the Cheyennes, made a speech full of insolent defiance.

"This is the first time," said the gigantic warrior, who was six feet three and magnificently proportioned,*

* General Fry, in his valuable book, "Army Sacrifices," now unfortunately out of print and scarce, thus graphically describes him: "A veritable man of war, the shock

CAPT. LOUIS H. CARPENTER — LIEUT. FREDK. H. BEECHER *
MAJ. GEORGE A. FORSYTH — SCOUT JACK STILLWELL

BEECHER'S ISLAND FIGHTERS

* Killed on the Island

All contemporary portraits except Stillwell's

"that I have ever shaken the white man's hand in friendship. If the railway is continued I shall be his enemy forever."

There was no stopping the railway. Its progress was as irresistible as the movement of civilization itself. The Indians went on the war-path. The Cheyennes were led by their two principal chiefs, Black Kettle being the second. We shall see subsequently how Custer accounted for Black Kettle. This story deals with the adventures of Roman Nose.

As fighters these Indians are entitled to every admiration. As marauders they merit nothing but censure. The Indians of the early days of the nation, when Pennsylvania and New York were border states, and across the Alleghenies lay the frontier, were cruel enough, as the chronicle of the times abundantly testify; but they were angels of light compared with the Sioux and Cheyennes, the Kiowas, Arapahoes and Comanches, and these in turn were almost admirable beside the Apache. The first-named group were as cruel as they knew how to be, and they did not lack knowledge, either. The Apaches were more ingenious and devilish in their practices than the others. The Sioux and the Cheyennes were brutal with the brutality of a wild bull or a grizzly bear. To that same kind of brutality the Apaches added the malignity of a wildcat and

of battle and scenes of carnage and cruelty were as of the breath of his nostrils; about thirty years of age, standing six feet three inches high, he towered giant-like above his companions. A grand head with strongly marked features, lighted by a pair of fierce black eyes; a large mouth with thin lips, through which gleamed rows of strong, white teeth; a Roman nose with dilated nostrils like those of a thoroughbred horse, first attracted attention, while a broad chest, with symmetrical limbs on which the muscles under the bronze of his skin stood out like twisted wire, were some of the points of this splendid animal. Clad in buckskin leggings and moccasins elaborately embroidered with beads and feathers, with a single eagle feather in his scalp-lock, and with that rarest of robes, a white buffalo, beautifully tanned and soft as cashmere, thrown over his naked shoulders, he stood forth, the war chief of the Cheyennes."

the subtlety of a snake. The men of the first group would stand out and fight in the open to gain their ends, although they did not prefer to. They were soldiers and warriors as well as torturers. The Apache was a lurking skulker, but, when cornered, a magnificent fighter also.* General Crook calls him "the tiger of the human species." However, from the point of detestableness there wasn't much to choose between them.

Perhaps we ought not to blame the Indians for acting just as our ancestors of, say the Stone Age, acted in all probability. And when you put modern weapons and modern whisky in the hands of the Stone Age men you need not be surprised at the consequences. The Indian question is a terrible one any way you take it. It cannot be denied they have been treated abominably by the United States, and that they have good cause for resentment; but the situation has been so peculiar that strife has been inevitable.

As patriots defending their country, they are not without certain definite claims to our respect. Recognizing the right of the aborigines to the soil, the government has yet arbitrarily abrogated that right at pleasure. At times the Indians have been regarded as independent nations, with which all differences were to be settled by treaty as between equals; and again, as a body of subjects whose affairs could be and would be administered willy-nilly by the United States. Such vacillations are certain to result in trouble, especially as, needless to say, the Indians invariably considered themselves as much independent nations as England and France might consider themselves, in dealing with the United States or with one another. And the Indians naturally

* Charles F. Lummis refers to the Apaches as among the most ferocious and most successful warriors in history.

claimed and insisted that the territory where their fathers had roamed for centuries belonged solely and wholly to them. They admitted no suzerainty of any sort, either. And they held the petty force the government put in the field in supreme contempt until they learned by bitter experience the illimitable power of the United States.

To settle such a growing question in a word, offhand, as it were, is, of course, impossible, nor does the settlement lie within the province of these articles; but it may be said that if the United States had definitely decided upon one policy or the other, and had then concentrated all its strength upon the problem; if it had realized the character of the people with whom it was dealing, and had made such display of its force as would have rendered it apparent, to the keenest as well as to the most stupid and besotted of the Indians, that resistance was entirely futile, things might have been different. But it is the solemn truth that never, in any of the Indian wars west of the Missouri, has there been a force of soldiers in the field adequate to deal with the question. The blood of thousands of soldiers and settlers — men, women, and children — might have been spared had this fact been realized and acted upon.

The Cheyennes swept through western Kansas like a devastating storm. In one month they cut off, killed, or captured eighty-four different settlers, including their wives and children. They swept the country bare. Again and again the different gangs of builders were wiped out, but the railroad went on. General Sheridan finally took the field in person, as usual with an inadequate force at his disposal. One of his aides-de-camp was a young cavalry officer named George Alexander Forsyth, commonly known to his friends as "Sandy"

Forsyth. He had entered the volunteer army in 1861 as a private of dragoons in a Chicago company. A mere boy, he had come out a brigadier-general. In the permanent establishment he was a major in the Ninth Cavalry. Sheridan knew him. He was one of the two officers who made that magnificent ride with the great commander that saved the day at Winchester, and it was due to his suggestion that Sheridan rode down the re-adjusted lines before they made the return advance which decided the fate of the battle. During all that mad gallop and hard fighting young Forsyth rode with the General. To-day he is the only survivor of that ride.

Forsyth was a fighter all through, and he wanted to get into the field in command of some of the troops operating directly on the Indians in the campaign under consideration. No officer was willing to surrender his command to Forsyth on the eve of active operations, and there was no way, apparently, by which he could do anything until Sheridan acceded to his importunities by authorizing him to raise a company of scouts for the campaign. He was directed, if he could do so, to enlist fifty men, who, as there was no provision for the employment of scouts or civilian auxiliaries, were of necessity carried on the payrolls as quartermasters' employees for the magnificent sum of one dollar per day. They were to provide their own horses, but were allowed thirty cents a day for the use of them, and the horses were to be paid for by the government if they were "expended" during the campaign. They were equipped with saddle, bridle, haversack, canteen, blanket, knife, tin cup, Spencer repeating rifle, good for seven shots without reloading, six in the magazine, one in the barrel, and a heavy Colt's army revolver. There were no tents or other similar conveniences, and four mules constituted the baggage train. The force was intended to be strictly mobile, and it was. Each man carried on his person one hundred and forty rounds of ammunition for his rifle and thirty rounds for his revolver. The four mules carried the medical supplies and four thousand rounds of extra ammunition. Each officer and man took seven days' rations. What he could not carry on his person was loaded on the pack mules; scanty rations they were, too.

As soon as it was known that the troop was to be organized, Forsyth was overwhelmed with applications from men who wished to join it. He had the pick of the frontier to select from. He chose thirty men at Fort Harker and the remaining twenty from Fort Hayes. Undoubtedly they were the best men in the West for the purpose. To assist him, Lieutenant Frederick H. Beecher, of the Third Infantry, was detailed as second in command. Beecher was a young officer with a record. He had displayed peculiar heroism at the great battle of Gettysburg, where he had been so badly wounded that he was lame for the balance of his life. He was a nephew of the great Henry Ward Beecher and a worthy representative of the distinguished family whose name he bore. The surgeon of the party was Dr. John H. Mooers, a highly-trained physician, who had come to the West in a spirit of restless adventure. He had settled at Hayes City and was familiar with the frontier. The guide of the party was Sharp Grover, one of the remarkable plainsmen of the time, regarded as the best scout in the government service. The first sergeant was W. H. H. McCall, formerly brigadier-general, United States Volunteers. McCall, in command of a Pennsylvania regiment, had been promoted for conspicuous gallantry on the field, when John B. Gordon made his magnificent dash out of Petersburg and attacked Fort Steadman.

The personnel of the troop was about equally divided between hunters and trappers and veterans of the Civil War, nearly all of whom had held commissions in either the Union or Confederate Army, for the command included men from both sides of Mason and Dixon's line. It was a hard-bitten, unruly group of fighters. Forsyth was just the man for them. While he did not attempt to enforce the discipline of the Regular Army, he kept them regularly in hand. He took just five days to get his men and start on the march. They left Fort Wallace, the temporary terminus of the Kansas Pacific Railroad, in response to a telegram from Sheridan that the Indians were in force in the vicinity, and scouted the country for some six days, finally striking the Indian trail, which grew larger and better defined as they pursued it. Although it was evident that the Indians they were chasing greatly outnumbered them, they had come out for a fight and wanted one, so they pressed on. They got one, too.*

II. The Island of Death

On the evening of the fifteenth of September, hot on the trail, now like a well-beaten road, they rode through a depression or a ravine, which gave entrance into a valley some two miles wide and about the same length. Through this valley ran a little river, the Arickaree. They encamped on the south bank of the river about

* The reason a large body of men had not been detailed for the pursuit was that the greater the number the slower the movement would have been, and the Indians could and would have kept out of the way with ease. If the Indians were laying a trap for Forsyth, he was tempting them to stop and fight.

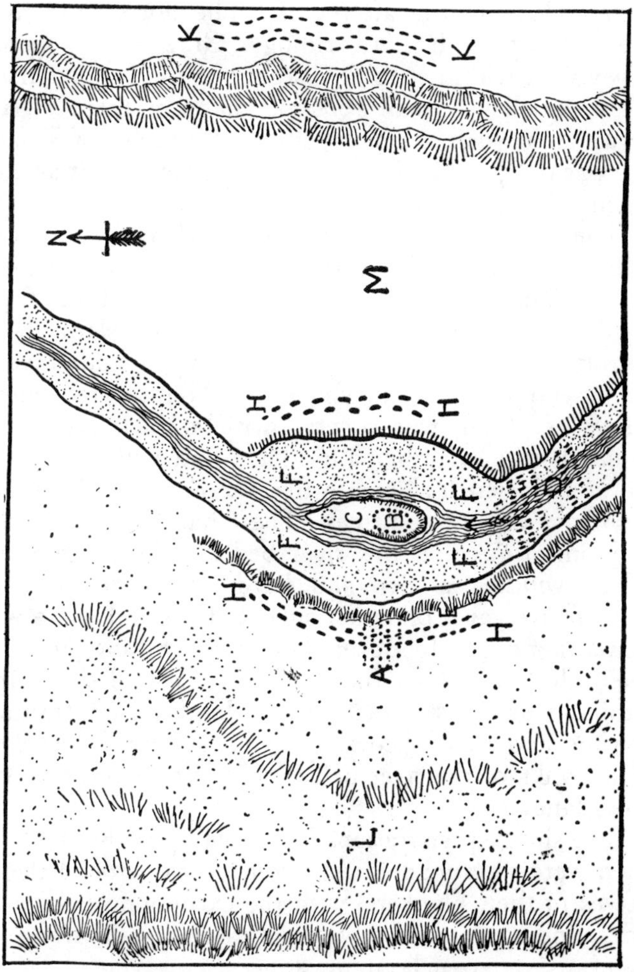

MAP OF FORSYTH'S DEFENSE OF BEECHER'S ISLAND, ARIKAREE RIVER, COLORADO

(*Drawn by the author from rough sketches and maps furnished by General Forsyth*)

Explanation of Map: A. Forsyth's camp before attack. B. Rifle-pits on island. C. Low, unoccupied land on island with solitary cottonwood at end. D. Indian charge led by Roman Nose and Medicine Man. EE. Low banks fringed with trees. FF. Dry sandy bed of the river. HH. Indian riflemen on the banks. KK. Indian women and children on bluffs, half a mile from river. L. Ground sloping gently to river, M. Level grassy plain to bluffs.

four o'clock in the afternoon. The horses and men were weary with hard riding. Grazing was good. They were within striking distance of the Indians now. Forsyth believed there were too many of them to run away from such a small body as his troop of scouts. He was right. The Indians had retreated as far as they intended to.

The river bed, which was bordered by wild plums, willows and alders, ran through the middle of the valley. The bed of the river was about one hundred and forty yards wide. In the middle of it was an island about twenty yards wide and sixty yards long. The gravelly upper end of the island, which rose about two feet above the water level, was covered with a thick growth of stunted bushes, principally alders and willows; at the lower end, which sloped to the water's edge, there rose a solitary cottonwood tree. There had been little rain for some time, and this river bed for the greater part of its width was dry and hard.* For a space of four or five yards on either side of the island there was water, not over a foot deep, languidly washing the gravelled shores. When the river bed was full the island probably was overflowed. Such islands form from time to time, and are washed away as quickly as they develop. The banks of the river bed on either side commanded the island.

The simple preparations for the camp of that body of men were soon made. As night fell they rolled themselves in their blankets, with the exception of the sentries, and went to sleep with the careless indifference of veterans under such circumstances.

* In dry seasons I have often seen Western river beds half a mile wide absolutely devoid of water. In the wet season these same beds would be roaring torrents from bank to bank.

BEECHER'S ISLAND FIELD

The battle took place just about where the cattle are standing in the river. The shifting current has obliterated the Island

Forsyth, however, as became a captain, was not so careless or so reckless as his men. They were alone in the heart of the Indian country, in close proximity to an overwhelming force, and liable to attack at any moment. He knew that their movements had been observed by the Indians during the past few days. Therefore the young commander was on the alert throughout the night, visiting the outposts from time to time to see that careful watch was kept.

Just as the first streaks of dawn began to "lace the severing clouds," he happened to be standing by the sentry farthest from the camp. Silhouetted against the sky-line they saw the feathered head of an Indian. For Forsyth to fire at him was the work of an instant. At the same time a party which had crept nearer to the picket line unobserved dashed boldly at the horses, and resorting to the usual devices with bells, horns, hideous yells, and waving buffalo robes, attempted to stampede the herd.

Men like those scouts under such circumstances slept with their boots on. The first shot called them into instant action. They ran instinctively to the picket line. A sharp fire, and the Indians were driven off at once. Only the pack mules got away. No pursuit was attempted, of course. Orders were given for the men to saddle their horses and stand by them. In a few moments the command was drawn up in line, each man standing by his horse's head, bridle reins through his left arm, his rifle grasped in his right hand — ready! Scarcely had the company been thus assembled when Grover caught Forsyth's arm and pointed down the valley.

"My God!" he cried, "look at the Injuns!"

In front of them, on the right of them, in the rear of

them, the hills and valleys on both sides of the river seemed suddenly to be alive with Indians. It was as quick a transformation from a scene of peaceful quiet to a valley filled with an armed force as the whistle of Roderick Dhu had effected in the Scottish glen.

The way to the left, by which they had entered the valley, was still open. Forsyth could have made a running fight for it and dashed for the gorge through which he had entered the valley. There were, apparently, no Indians barring the way in that direction. But Forsyth realized instantly that for him to retreat would mean the destruction of his command, that the Indians had in all probability purposely left him that way of escape, and if he tried it he would be ambushed in the defile and slain. That was just what they wanted him to do, it was evident. That was why he did not attempt it. He was cornered, but he was not beaten, and he did not think he could be. Besides, he had come for that fight, and that fight he was bound to have.

Whatever he was to do he must do quickly. There was no place to which he could go save the island. That was not much of a place at best, but it was the one strategic point presented by the situation. Pouring a heavy fire into the Indians, Forsyth directed his men to take possession of the island under cover of the smoke. In the movement everything had to be abandoned, including the medical stores and rations, but the precious ammunition — that must be secured at all hazards. Protected by a squad of expert riflemen on the river bank, who presently joined them, the scouts reached the island in safety, tied their horses to the bushes around the edge of it, and in the intervals of fighting set to work digging rifle-pits covering an ellipse twenty by forty yards, one pit for each man, with which to defend the upper and higher part of the island They had nothing to dig with except tin cups, tin plates, and their bowie knives, but they dug like men. There was no lingering or hesitation about it.

The chief of the Indian force, which was made up of Northern Cheyennes, Oglala and Brulé Sioux, with a few Arapahoes and a number of Dog Soldiers, was the famous Roman Nose, an enemy to be feared indeed. He was filled with disgust and indignation at the failure of his men to occupy the island, the strategic importance of which he at once detected. It is believed that orders to seize the island had been given, but for some reason they had not been obeyed; and to this oversight or failure was due the ultimate safety of Forsyth's men. It was not safe to neglect the smallest point in fighting with a soldier like Forsyth.

With more military skill than they had ever displayed before, the Indians deliberately made preparations for battle. The force at the disposal of Roman Nose was something less than one thousand warriors. They were accompanied by their squaws and children. The latter took position on the bluffs on the east bank of the river, just out of range, where they could see the whole affair. Like the ladies of the ancient tournaments, they were eager to witness the fighting and welcome the victors, who, for they never doubted the outcome, were certain to be their own.

Roman Nose next lined the banks of the river on both sides with dismounted riflemen, skilfully using such concealment as the ground afforded. The banks were slightly higher than the island, and the Indians had a plunging fire upon the little party. The riflemen on the banks opened fire at once. A storm of bullets was poured upon the devoted band on the island. The scouts, husbanding their ammunition, slowly and deliberately replied, endeavoring, with signal success, to make every shot tell. As one man said, they reckoned "every ca'tridge was wuth at least one Injun." The horses of the troop, having no protection, received the brunt of the first fire. They fell rapidly, and their carcasses rising in front of the rifle-pits afforded added protection to the soldiers. There must have been a renegade white man among the savages, for in a lull of the firing the men on the island heard a voice announce in perfect English, "There goes the last of their horses, anyway." Besides this, from time to time, the notes of an artillery bugle were heard from the shore. The casualties had not been serious while the horses stood, but as soon as they were all down the men began to suffer.*

During this time Forsyth had been walking about in the little circle of defenders encouraging his men. He was met on all sides with insistent demands that he lie down and take cover, and, the firing becoming hotter, he at last complied. The rifle-pit which Surgeon Mooers had made was a little wider than that of the other men, and as it was a good place from which to direct the fighting, at the doctor's suggestion some of the scouts scooped it out to make it a little larger, and Forsyth lay down by him.

The fire of the Indians had been increasing. Several scouts were killed, more mortally wounded, and some slightly wounded. Doctor Mooers was hit in the forehead and mortally wounded. He lingered for three days, saying but one intelligent word during the whole period. Although he was blind and speechless, his motions some-

* As the Indians surrounded the island and the fire came in from all quarters, the men had to dig the earth for protection in rear as well as in front, and the rifle-pits were, in fact, hollows scooped out of the ground just long enough for a man to lie in.

times indicated that he knew where he was. He would frequently reach out his foot and touch Forsyth. A bullet struck Forsyth in the right thigh, and glancing upwards bedded itself in the flesh, causing excruciating pain. He suffered exquisite anguish, but his present sufferings were just beginning, for a second bullet struck him in the leg, between the knee and ankle, and smashed the bone, and a third glanced across his forehead, slightly fracturing his skull and giving him a splitting headache, although he had no time to attend to it then.

III. The Charge of the Five Hundred

During all this time Roman Nose and his horsemen had withdrawn around the bend up the river, which screened them from the island. At this juncture they appeared in full force, trotting up the bed of the river in open order in eight ranks of about sixty front. Ahead of them, on a magnificent chestnut horse, trotted Roman Nose. The warriors were hideously painted, and all were naked except for moccasins and cartridge belts. Eagle feathers were stuck in their long hair, and many of them wore gorgeous feather war bonnets. They sat their horses without saddles or stirrups, some of them having lariats twisted around the horses' bellies like a surcingle. Roman Nose wore a magnificent war bonnet of feathers streaming behind him in the wind and surmounted by two buffalo-horns; around his waist he had tied an officer's brilliant scarlet silk sash, which had been presented to him at the Fort Ellsworth conference. The sunlight illumined the bronze body of the savage Hercules, exhibiting the magnificent proportions of the man. Those who followed him were in every way worthy of their leader.

As the Indian cavalry appeared around the bend to the music of that bugle, the fire upon the island from the banks redoubled in intensity. Forsyth instantly divined that Roman Nose was about to attempt to ride him down. He also realized that, so soon as the horses were upon him, the rifle fire from the bank would of necessity be stopped. His order to his men was to cease firing, therefore; to load the magazines of their rifles, charge their revolvers, and wait until he gave the order to fire. The rifles of the dead and those of the party too severely wounded to use them were distributed among those scouts yet unharmed. Some of the wounded insisted upon fighting. Forsyth propped himself up in his rifle-pit, his back and shoulders resting against the pile of earth, his rifle and revolver in hand. He could see his own men, and also the Indians coming up the river.

Presently, shouting their war songs, at a wild pealed whoop from their chief, the Indian horsemen broke into a gallop, Roman Nose leading the advance, shaking his heavy Spencer rifle — captured, possibly, from Fetterman's men — in the air as if it had been a reed. There was a last burst of rifle fire from the banks, and the rattle of musketry was displaced by the war songs of the Indians and the yells of the squaws and children on the slopes of the hills. As the smoke drifted away on that sunny September morning, they saw the Indians almost upon them. In spite of his terrible wounds the heroic Forsyth was thoroughly in command. Waiting until the tactical moment when the Indians were but fifty yards away and coming at a terrific speed, he raised himself on his hands to a sitting position and cried, "Now!"

The men rose to their knees, brought their guns to

ROMAN NOSE LEADING THE CHARGE AGAINST FORSYTH'S DEVOTED BAND

Drawing by Charles Schreyvogel

their shoulders, and poured a volley right into the face of the furious advance. An instant later, with another cartridge in the barrel they delivered a second volley. Horses and men went down in every direction; but, like the magnificent warriors they were, the Indians closed up and came sweeping down. The third volley was poured into them. Still they came. The war songs had ceased by this time, but in undaunted spirit, still pealing his war cry above the crashing of the bullets, at the head of his band, with his magnificent determination unshaken, Roman Nose led such a ride as no Indian ever attempted before or since. And still those quiet, cool men continued to pump bullets into the horde. At the fourth volley the medicine man on the left of the line and the second in command went down. The Indians hesitated at this reverse, but swinging his rifle high in the air in battle frenzy, the great war chief rallied them, and they once more advanced. The fifth volley staggered them still more. Great gaps were opened in their ranks. Horses and men fell dead, but the impetus was so great, and the courage and example of their leader so splendid, that the survivors came on unchecked. The sixth volley did the work. Just as he was about to leap on the island, Roman Nose and his horse were both shot to pieces. The force of the charge, however, was so great that the line was not yet entirely broken. The horsemen were within a few feet of the scouts, when the seventh volley was poured into their very faces. As a gigantic wave meets a sharply jutting rock and is parted, falling harmlessly on either side of it, so was that charge divided, the Indians swinging themselves to the sides of their horses as they swept down the length of the island.

The scouts sprang to their feet at this juncture, and

almost at contact range jammed their revolver shots at the disorganized masses. The Indians fled precipitately to the banks on either side, and the yelling of the war chants of the squaws and children changed into wails of anguish and despair, as they marked the death of Roman Nose and the horrible slaughter of his followers.

It was a most magnificent charge, and one which for splendid daring and reckless heroism would have done credit to the best troops of any nation in the world. And magnificently had it been met. Powell's defense of the corral on Piney Island was a remarkable achievement, but it was not to be compared to the fighting of these scouts on the little open, unprotected heap of sand and gravel in the Arickaree.

As soon as the Indian horsemen withdrew, baffled and furious, a rifle fire opened once more from the banks. Lieutenant Beecher, who had heroically performed his part in the defense, crawled over to Forsyth and said:

"I have my death wound, General. I am shot in the side and dying."

He said the words quietly and simply, as if his communication was utterly commonplace, then stretched himself out by his wounded commander, lying, like Steerforth, with his face upon his arm.

"No, Beecher, no," said Forsyth, out of his own anguish; "it can not be as bad as that."

"Yes," said the young officer, "good-night."

There was nothing to be done for him. Forsyth heard him whisper a word or two of his mother, and then delirium supervened. By evening he was dead. In memory of the brave young officer, they called the place where he had died Beecher's Island.

At two o'clock in the afternoon a second charge of horse was assayed in much the same way as the first had been delivered; but there was no longer a great war chief in command, and this time the Indians broke at one hundred yards from the island. At six o'clock at night they made a final attempt. The whole party, horse and foot, in a solid mass rushed from all sides upon the island. They came forward, yelling and firing, but they were met with so severe a fire from the rifle-pits that, although some of them actually reached the foot of the island, they could not maintain their position, and were driven back with frightful loss. The men on the island deliberately picked off Indian after Indian as they came, so that the dry river bed ran with blood. The place was a very hell to the Indians. They withdrew at last, baffled, crushed, beaten.

With nightfall the men on the island could take account of the situation. Two officers and four men were dead or dying, one officer and eight men were so severely wounded that their condition was critical. Eight men were less severely wounded, making twenty-three casualties out of fifty-one officers and men.* There were no rations, but thank God there was an abundance of water. They could get it easily by digging in the sandy surface of the island. They could subsist, if necessary, on strips of meat cut from the bodies of the horses. The most serious lack was of medical attention. The doctor lying unconscious, the wounded were forced to get along with the unskilled care of their comrades, and with water, and rags torn from clothing for dressings. Little could be done for them. The day had been frightfully hot, but, fortunately, a heavy rain

* Two of the scouts had been left behind, at Fort Wallace, because of illness.

fell in the night, which somewhat refreshed them. The rifle-pits were deepened and made continuous by piling saddles and equipments, and by further digging in the interspaces.

One of the curious Indian superstitions, which has often served the white man against whom he has fought to good purpose, is that when a man is killed in the dark he must pass all eternity in darkness. Consequently, he rarely ever attacks at night. Forsyth's party felt reasonably secure from any further attack, therefore, notwithstanding which they kept watch.

IV. The Siege of the Island

As soon as darkness settled down volunteers were called for to carry the news of their predicament to Fort Wallace, one hundred miles away. Every man able to travel offered himself for the perilous journey. Forsyth selected Trudeau and Stillwell. Trudeau was a veteran hunter, Stillwell a youngster only nineteen years of age, although he already gave promise of the fame as a scout which he afterwards acquired. To them he gave the only map he possessed. They were to ask the commander of Fort Wallace to come to his assistance. As soon as the two brave scouts had left, every one realized that a long wait would be entailed upon the little band, if, indeed, it was not overwhelmed meanwhile, before any relieving force could reach the island. And there were grave doubts as to whether, in any event, Trudeau and Stillwell could get through the Indians. It was not a pleasant night they spent, therefore, although they were busy strengthening the defenses, and nobody got any sleep.

Early the next morning the Indians again made their

appearance. They had hoped that Forsyth and his men would have endeavored to retreat during the night, in which event they would have followed the trail and speedily annihilated the whole command. But Forsyth was too good a soldier to leave the position he had chosen. During the fighting of the day before he had asked Grover his opinion as to whether the Indians could deliver any more formidable attack than the one which had resulted in the death of Roman Nose, and Grover, who had had large experience, assured him that they had done the best they could, and indeed better than he or any other scout had ever seen or heard of in any Indian warfare. Forsyth was satisfied, therefore, that they could maintain the position, at least until they starved.

The Indians were quickly apprised, by a volley which killed at least one man, that the defenders of the island were still there. The place was closely invested, and although the Indians made several attempts to approach it under a white flag, they were forced back by the accurate fire of the scouts, and compelled to keep their distance. It was very hot. The sufferings of the wounded were something frightful. The Indians were having troubles of their own, too. All night and all day the defenders could hear the beating of the tom-toms or drums and the mournful death songs and wails of the women over the bodies of the slain, all but three of whom had been removed during the night.* These three were

* The reason why an Indian will sacrifice everything to remove the body of one of his tribe or kin who has been killed, is to prevent the taking of his scalp. The religious belief of the Indians is that a man who is scalped cannot enter the happy hunting grounds, but is doomed to wander in outer darkness forever. For that reason he always scalps his enemy, so that when he himself reaches the happy hunting grounds he will not be bothered by a lot of enemies whom he has met and overcome during his lifetime. Naturally, it was a point of honor for him to get the bodies of his friends away, so that they might not be debarred from the Indian Heaven in the hereafter. Sometimes, however, the In-

lying so near the rifle-pits that the Indians did not dare to approach near enough to get them. The three dead men had actually gained the shore of the island before they had been killed.

The command on the island had plenty to eat, such as it was. There was horse and mule meat in abundance. They ate it raw, when they got hungry enough. Water was plentiful. All they had to do was to dig the rifle-pits a little deeper, and it came forth in great quantities. It was weary waiting, but there was nothing else to do. They dared not relax their vigilance a moment. The next night, the second, Forsyth despatched two more scouts, fearing the first two might not have got through, thus seeking to "make assurance double sure." This pair was not so successful as the first. They came back about three o'clock in the morning, having been unable to pass the Indians, for every outlet was heavily guarded.

The third day the Indian women and children were observed withdrawing from the vicinity. This cheered the men greatly, as it was a sign that the Indians intended to abandon the siege. The warriors still remained, however, and any incautious exposure was a signal for a volley. That night two more men were despatched with an urgent appeal, and these two succeeded in getting through. They bore this message:

dian did not scalp the body of a particularly brave man, for this reason: It is his belief that if he kills a man in battle and does not scalp him, that man will be his slave or servant in the happy hunting grounds, and although the victim still possesses capacities for mischief, the Indian sometimes risks all in the future glory that will come to him from holding in slavery a brave man, or a noted warrior, as a spiritual witness to his prowess. It is stated that the Indians never scalp the bodies of negroes and suicides. "Buffalo soldier heap bad medicine," is their universal testimony when asked why they do not scalp negro troopers whom they have killed or captured. Perhaps they cannot scalp a woolly, kinky-haired black soldier, and that is the reason it is "bad medicine." Suicide is "bad medicine," too, for some unexplained reason.

"Sept. 19, 1868.

To Colonel Bankhead, or Commanding Officer,
Fort Wallace:

I sent you two messengers on the night of the 17th inst., informing you of my critical condition. I tried to send two more last night, but they did not succeed in passing the Indian pickets, and returned. If the others have not arrived, then hasten at once to my assistance. I have eight badly wounded and ten slightly wounded men to take in. . . . Lieutenant Beecher is dead, and Acting Assistant Surgeon Mooers probably cannot live the night out. He was hit in the head Thursday, and has spoken but one rational word since. I am wounded in two places — in the right thigh, and my left leg is broken below the knee. . . .

I am on a little island, and have still plenty of ammunition left. We are living on mule and horse meat, and are entirely out of rations. If it was not for so many wounded, I would come in, and take the chances of whipping them if attacked. They are evidently sick of their bargain. . . . I can hold out for six days longer if absolutely necessary, but please lose no time.

P. S. — My surgeon having been mortally wounded, none of my wounded have had their wounds dressed yet, so please bring out a surgeon with you."

The fourth day passed like the preceding, the squaws all gone, the Indians still watchful. The wound in Forsyth's leg had become excruciatingly painful, and he begged some of the men to cut out the bullet. But they discovered that it had lodged near the femoral artery, and fearful lest they should cut the artery and the young commander should bleed to death, they positively refused. In desperation, Forsyth cut it out himself.

He had his razor in his saddle bags and, while two men pressed the flesh back, he performed the operation successfully, to his immediate relief.

The fifth day the mule and horse meat became putrid and therefore unfit to eat. An unlucky coyote wandered over to the island, however, and one of the men was fortunate enough to shoot him. Small though he was, he was a welcome addition to their larder, for he was fresh. There was but little skirmishing on the fifth day, and the place appeared to be deserted. Forsyth had half a dozen of his men raise him on a blanket above the level of the rifle beds so that he might survey the scene himself. Not all the Indians were gone, for a sudden fusillade burst out from the bank. One of the men let go the corner of the blanket which he held while the others were easing Forsyth down, and he fell upon his wounded leg with so much force that the bone protruded through the flesh. He records that he used some severe language to that scout.

On the sixth day Forsyth assembled his men about him, and told them that those who were well enough to leave the island would better do so and make for Fort Wallace; that it was more than possible that none of the messengers had succeeded in getting through; that the men had stood by him heroically, and that they would all starve to death where they were unless relief should come; and that they were entitled to a chance for their lives. He believed the Indians, who had at last disappeared, had received such a severe lesson that they would not attack again, and that if the men were circumspect they could get through to Fort Wallace in safety. The wounded, including himself, must be left to take care of themselves and take the chances of escape from the island.

The proposition was received in surprised silence for a few moments, and then there was a simultaneous shout of refusal from every man: "Never! We'll stand by you." McCall, the first sergeant and Forsyth's right-hand man since Beecher had been killed, shouted out emphatically: "We've fought together, and, by Heaven, if need be, we'll die together."

They could not carry the wounded; they would not abandon them. Remember these men were not regular soldiers. They were simply a company of scouts, more or less loosely bound together, but, as McCall had pointed out, they were tied to one another by something stronger than discipline. Not a man left the island, although it would have been easy for the unwounded to do so, and possibly they might have escaped in safety.

For two more days they stood it out. There was no fighting during this time, but the presence of an Indian vedette indicated that they were under observation. They gathered some wild plums and made some jelly for the wounded; but no game came their way, and there was little for them to do but draw in their belts a little tighter and go hungry, or, better, go hungrier. On the morning of the ninth day, one of the men on watch suddenly sprang to his feet, shouting:

"There are moving men on the hills." Everybody who could stand was up in an instant, and Grover, the keen-eyed scout, shouted triumphantly:

"By the God above us, there's an ambulance!" They were rescued at last.

NOTE.—The serial publication of this article called forth another version of this affair, differing from it in some non-essential features, which was written by Mr. Herbert Myrick, and published serially. Mr. Myrick accounts for the "mysterious voice" which the scouts heard saying in English, "There goes the last of their horses anyway," by disclosing the interesting fact that there were two renegade white men among the Indians. One of them was called "Nibsi" or "Black Jack," a notorious desperado, who was afterwards hung for murder. The other was Jack Clybor, once a trooper of the Seventh Cavalry. Having been shot and left for dead in an engagement, the Indians captured him, nursed him back to life, adopted him, and named him "Comanche." He was a singular compound of good and evil, and became as notorious for his good deeds as for his bad acts. Mr. Myrick has been collecting a mass of unknown and unpublished Western material for many years, which when published will undoubtedly clear up many mysteries, throw light upon many disputed questions, and prove of the deepest interest as well.

CHAPTER SIX

The Journey of the Scouts and the Rescue of Forsyth

I. The Adventures of the Scouts

TRUDEAU and Stillwell, the first pair of scouts despatched by Forsyth with the story of his desperate situation on Beecher's Island, left their commander about midnight on the evening of the first day of the attack. The Indians had withdrawn from the immediate vicinity of the river and were resting quietly in the camps on either side, although there were a number of warriors watching the island. The men bade a hasty good-by to their comrades, received their captain's final instructions, and with beating hearts stole away on their desperate errand.

They neglected no precaution that experience could dictate. They even took off their boots, tied them together by the straps, slung them around their necks, and walked backward down the bed of the river in their stocking feet, so that, if the Indians by any chance stumbled upon their trail the next morning, it would appear to have been made by moccasined feet and perhaps escape attention, especially as the tracks would point toward the island instead of away from it. Further to disguise themselves, they wrapped themselves in blankets, which they endeavored to wear as the Indians did.

They proceeded with the most fearsome caution. Such was the circumspection with which they moved and the care necessary because of the watchfulness of the foe, who might be heard from time to time moving about on the banks, that by daylight they had progressed but two miles. During most of the time after leaving the river bed they had crawled on their hands and knees. Before sunrise they were forced to seek such concealment as they could find in a washout, a dry ravine, within sight and sound of the Indian camps. Providence certainly protected them, for if any of the Indians had happened to wander in their direction there was nothing to prevent their discovery; and if the savages had stumbled upon their hiding-place it would have been all up with them. Death by torture would have been inevitable if they were taken alive, and the only way to prevent that would be suicide. They had determined upon that. They had pledged each other to fight until the last cartridge, and to save that for themselves. They had nothing to eat and nothing to drink. The sun beat down upon them fiercely all the long day. After their experience of the one before, it was a day calculated to break down the strongest of men. They bore up under the strain, however, as best they could, and when darkness came they started out once more.

This night there was no necessity for so much caution and they made better progress, although they saw and successfully avoided several parties of Indians. When the day broke they were forced to conceal themselves again. The country was covered with wandering war parties, and it was not yet safe to travel by daylight.

"SIMULTANEOUSLY WITH THEIR ARRIVAL A RATTLESNAKE MADE HIS APPEARANCE"

Drawing by Will Crawford

This day they hid themselves under the high banks of a river. Again they were fortunate in remaining unobserved, although several times bands of warriors passed near them. They traveled all the third night, making great progress. Morning found them on an open plain with no place to hide in but a buffalo wallow — a dry alkali mud-hole which had been much frequented in the wet season by buffalo — which afforded scanty cover at best.

During this day a large party of scouting Indians halted within one hundred feet of the wallow. Simultaneously with their arrival a wandering rattlesnake made his appearance in front of the two scouts, who were hugging the earth and expecting every minute to be discovered. The rattlesnake in his way was as deadly as the Indians. The scouts could have killed him easily had it not been for the proximity of the Cheyennes. To make the slightest movement would call attention to their hiding-place. Indeed, the sinister rattle of the venomous snake before he struck would probably attract the notice of the alert Indians. Between the savage reptile and the savage men the scouts were in a frightful predicament, which young Stillwell, a lad of amazing resourcefulness, instantly and effectually solved. He was chewing tobacco at the time, and as the snake drew near him and made ready to strike, he completely routed him by spitting tobacco juice in his mouth and eyes and all over his head. The rattlesnake fled; he could not stand such a dose. The Indians presently moved on, having noticed nothing, and so ended perhaps the most terrible half hour the two men had ever experienced.

They started early on the evening of the fourth night, and this time made remarkable progress. Toward morning, however, Trudeau all but broke down. The brunt of the whole adventure thereupon fell on Stillwell. He encouraged his older companion, helped him along as best he could, and finally, late at night, they reached Fort Wallace and told their tale. Instantly all was excitement in the post. Captain and Brevet Lieutenant-Colonel Louis H. Carpenter, with seventy men of Troop H, of the Tenth Cavalry (a negro regiment), with Lieutenants Banzhaf and Orleman, Doctor Fitzgerald and seventeen scouts, with thirteen wagons and an ambulance, had been sent out from the post the day before with orders to make a camp on the Denver road, about sixty miles from the fort. From there he was to scout in every direction, keep off the Indians, and protect trains.

At eleven o'clock at night a courier was despatched to Carpenter with the following order:

"Headquarters, Fort Wallace, Kansas,
September 22, 1868, 11:00 P. M.

Brevet Lieut.-Colonel L. H. CARPENTER, 10th U. S. Cavalry. On Scout.

Colonel:

The Commanding Officer directs you to proceed at once to a point on the "Dry Fork of the Republican," about seventy-five or eighty miles north, northwest from this point, thirty or forty miles west by a little south from the forks of the Republic, with all possible despatch.

Two scouts from Colonel Forsyth's command arrived here this evening and bring word that he (Forsyth) was attacked on the morning of Thursday last by an overwhelming force of Indians (700), who killed all the animals, broke Colonel Forsyth's left leg with a rifle ball, severely wounding him in the groin, wounded Doctor Mooers in the head, and wounded Lieutenant Beecher in several places. His back is supposed to be broken. Two men of the command were killed and eighteen or twenty wounded.

The men bringing the word crawled on hands and knees two miles, and then traveled only by night on account of the Indians, whom they saw daily.

Forsyth's men were intrenched in the dry bed of the creek with a well in the trench, but had only horse-flesh to eat and only sixty rounds of ammunition.

General Sheridan orders that the greatest despatch be used and every means employed to succor Forsyth at once. Colonel Bradley with six companies is now supposed by General Sheridan to be at the forks of the Republic.

Colonel Bankhead will leave here in one hour with one hundred men and two mountain howitzers.

Bring all your scouts with you.

Order Doctor Fitzgerald at once to this post, to replace Doctor Turner, who accompanies Colonel Bankhead for the purpose of dressing the wounded of Forsyth's party.

I am, Colonel, very respectfully your obedient servant,

HUGH JOHNSON,
1st Lieutenant 5th Infantry." Acting Post Adjutant.

One hour afterward Bankhead himself, with one hundred men and two howitzers and the surgeon, started for the relief of Forsyth. With Bankhead went the undaunted Stillwell as guide. Trudeau had suffered so much during the perilous journey that he was unable to accompany the relief party, and he soon afterward died from the hardships and excitement of the horrible days he had passed through.

II. The Rescue of Forsyth

Carpenter had bivouacked on the evening of the 22d of September at Cheyenne Wells, about thirty-five miles from Fort Wallace. He had broken camp early in the morning and had marched some ten miles, when, from a high point on a divide he had reached, which permitted a full view of the Rocky Mountains from Pike's to Long's Peaks, he observed a horseman galloping frantically toward them. He was the courier despatched by Colonel Bankhead. Carpenter was a splendid soldier. He had received no less than four brevets for gallantry during the Civil War. He had been on Sheridan's staff with Forsyth, and the two were bosom friends. No task could have been more congenial to him than this attempt at rescue.

He communicated the situation of their white comrades to his black troopers, and their officers crowded close about him. The orders were received with exultant cheers. The regiment had been raised since the war, and had not yet had a chance to prove its mettle. There were no veterans among them, and Carpenter and the other officers had been obliged to build the regiment from the ground up. Now was an opportunity to show what they could do. Carpenter had been trained to obey orders to the letter. In this instance he determined to disobey the command regarding Doctor Fitzgerald. It appeared to him that Bankhead had little hope that he (Carpenter) would find Forsyth, for he had sent him no guide; but Carpenter perceived that if he did find Forsyth — and he intended to find him — the conditions would be such that the services of a physician would be vitally necessary. He therefore retained the doctor. He also retained the wagon train, having no other way of carrying necessary supplies. For one

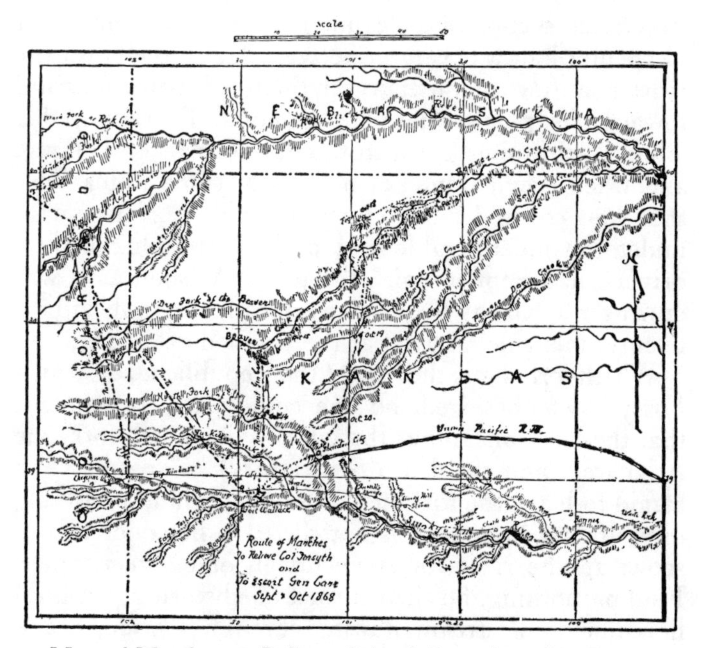

Map of Marches to Relieve Colonel Forsyth and to Escort General Carr, Drawn by General Carpenter

reason, if he had detached a guard for the train, it would have weakened his force so greatly as to have made it inadequate to the enterprise. The mules were strong and fresh, and he decided to keep the wagons with him. The pace was to be a fast one, and he instructed the wagon masters that, if any of the mule teams gave out, they should be shot and, if necessary, the wagon should be abandoned.

There was no one in his command, he found, who had ever been in that territory. Indeed, it is probable that, save Forsyth's men, no white men had ever penetrated that section of the country before. The map that Carpenter had was very defective. He studied over the matter a few moments, and then led his command toward the place where he supposed Forsyth to be. They advanced at a fast trot, with intervals of walking, and when they camped at night near some water holes they had covered nearly forty-five miles. The mules, under the indefatigable and profane stimulus of their drivers, had kept up with the rest. As soon as it was dawn the next day they started once more, and, after a twenty-mile ride, arrived at the dry bed of a river.

Whether this was the fork of the Republican, on which Forsyth was besieged, no one could tell. It happens that the Republican has three forks — a north fork, the Arickaree, and the south or dry fork. Carpenter was afraid to leave the fork he had found without satisfying himself that Forsyth was not there, so he concluded to scout up the river for some fifteen or eighteen miles. Finding nothing, he then turned northward again until he came to a stream flowing through a wide, grass-covered valley surrounded by high hills. As they entered the valley they came across a very large, fresh Indian trail. The scouts estimated that at least two thousand ponies had passed along the trail within a few hours. Various other signs showed a large village had moved down the trail.

They had traveled over forty miles this second day, and were apprehensive that the Indians, being so close to them, might attack them. It was nearly evening. A spot well adapted for defense was chosen near the water, the wagons were corralled, and preparations made for a stout resistance in case of an attack. While the men were making camp, Carpenter with a small escort rode to the top of one of the high hills bordering the valley. He could see for miles, but discovered no Indians nor any other living object in any direction. In front of them, however, on the top of another hill, were a number of scaffolds, each one bearing a human body. The Cheyenne method of burial was instantly recognized. A nearer look developed that the scaffolds had been recently erected. Five of them were examined, and in each case the body contained was that of a Cheyenne warrior, who had been killed by a gunshot wound. This was proof positive that they were some of the Indians who had been fighting against Forsyth.

While this was going on, one of the troopers noticed something white in a ravine on the opposite side of the valley. They galloped over to it, and found it to be an elaborate and beautiful tepee or wigwam, made out of freshly tanned white buffalo skins. The colonel dismounted, opened the tepee, and entered. There, upon a brush heap, lay a human figure wrapped in buffalo robes. When the robes were taken away the body of a splendid specimen of Indian manhood was disclosed. "He lay like a warrior taking his rest, with his martial cloak around him." His stern and royal look, the iron majesty of his features, even though composed in death, revealed at once a native chieftain. In his breast was a great, gaping wound, which had pierced his heart. He lay in his war-gear, with his weapons and other personal property close at hand.

After the examination they re-covered him and left him undisturbed. Then they went back to the camp.

The corral was watchfully guarded during the night, but no one appeared to molest them. It was decided to follow the Indian trail at daylight, as it would probably lead to the site of Forsyth's fight. Early the next morning, while they were packing up, they saw some horsemen coming over the hills to the south of them. They were white men, led by a scout named Donovan. Two more men had been despatched by Forsyth from the island on the third night of the siege, and being unobserved by the Indians, they had made their way to Fort Wallace. When they arrived there they found that Colonel Bankhead had already gone; whereupon Donovan had assembled five bold spirits and had immediately started out on the return journey. Fortunately for Carpenter, Donovan had struck the latter's trail, and had followed it to the camp.

Carpenter thereupon took thirty of his best mounted troopers and the ambulance loaded with hard-tack, coffee, and bacon, and set out on a gallop in the direction in which they supposed the island lay. Banzhaf was left in command of the rest, with orders to come on as fast as he could.

Carpenter went forward at a rapid gallop, and after traveling eighteen miles, while it was yet early in the morning, came to a spur of land from which he had a view of the surrounding country for miles. As he checked his horse on the brink, he saw to the right of him a valley through which meandered a narrow silver stream.

In the center of the valley there was an island. From it rose a solitary cottonwood. Men could be seen moving about the place. Donovan recognized it instantly. The horses of the detachment were put to a run, and the whole party galloped down the valley toward the island.

The scouts swarmed across the river with cries of joy, and welcomed the soldiers. The faithful mules dragged the ambulance close behind. There was food for everybody. Carpenter was struck with the wolfish look on the faces of the hungry men as they crowded around the ambulance. Later one of them brought him a piece of mule or horse meat which was to have been served for dinner that day, if the rescuers had not appeared. Carpenter could not endure even the odor of it.

Galloping across the river bed, the first to enter the rifle-pits on the island was Carpenter. There, on the ground before him, lay Forsyth. And what do you suppose he was doing? He was reading a novel! Some one had found, in an empty saddle-bag, an old copy of *Oliver Twist.* Forsyth was afraid to trust himself. He was fearful that he would break down. He did not dare look at Carpenter or express his feelings. Therefore he made a pretense of being absorbed in his book.

The black cavalry had arrived in the very nick of time. Forsyth was in a burning fever. Blood-poisoning had set in, and his wounds were in a frightful condition. Another day and it would have been too late. Everything was gone from him but his indomitable resolution. Many others were in like circumstances. It was well that Carpenter had brought his surgeon with him, for his services were sadly needed. The men were taken off the island, moved half a mile away from the terrible stench arising from the dead animals; the wagon train came up, camps were made, the dead were buried on the island they had immortalized with their valor, and everything possible done for the comfort of the living by their negro comrades.

The doctor wanted to amputate Forsyth's leg, but he protested, so that the amputation was not performed, and the leg was finally saved to its owner. One of the scouts, named Farley, however, was so desperately wounded that amputation had to be resorted to. The doctor performed the operation, assisted by Carpenter. A military commander in the field has to do a great many things.

The next day Bankhead made his appearance with his detachment. He had marched to the forks of the river and followed the Arickaree fork to the place. He was accompanied by two troops of the Second Cavalry, picked up on the way. He did not find fault with Carpenter for his disobedience in retaining Doctor Fitzgerald. On the contrary, such was his delight at the rescue that he fairly hugged his gallant subordinate.

As soon as it was possible, the survivors were taken back to Fort Wallace. Forsyth and the more severely wounded were carried in the ambulance. It took four days to reach the fort. Their progress was one long torture, in spite of every care that could be bestowed upon them. There was no road, and while the drivers chose the best spots on the prairie, there was, nevertheless, an awful amount of jolting and bumping.

Forsyth was brevetted a brigadier-general in the Regular Army for his conduct in this action. This was some compensation for two years of subsequent suffering until his wounds finally healed.

III. The End of Roman Nose

On the way back the men stopped at the white tepee in the lonely valley. Grover and McCall rode over to the spot with the officers and examined the body of the chieftain. They instantly identified him as Roman

Nose. With a touch of sentiment unusual in frontiersmen they respected his grave, and for the sake of his valor allowed him to sleep on undisturbed. His arms and equipments, however, were considered legitimate spoils of war, and were taken from him. It was a sad end, indeed, to all his splendid courage and glorious defiance of his white foemen.

The loss of the Indians in the several attacks was never definitely ascertained. They admitted to seventy-five killed outright and over two hundred seriously wounded, but it is certain that their total losses were much greater. The fighting was of the closest and fiercest description, and the Indians were under the fire of one of the most expert bodies of marksmen on the plains at half pistol-shot distance in the unique and celebrated battle. The whole action is almost unparalleled in the history of our Indian wars, both for the thrilling and gallant cavalry charge of the Indians and the desperate valor of Forsyth and his scouts.

IV. A Few Words About Forsyth's Men

The heroism and pluck of the men in the fight had been quite up to the mark set by their captain. A man named Farley had fought through the action with a severe bullet wound in the shoulder, which he never mentioned until nightfall; his father was mortally wounded, but he lay on his side and fought through the whole of the long first day until he died. Another man named Harrington was struck in the forehead by an arrow. He pulled out the shaft, but the head remained imbedded in the bone. An Indian bullet struck him a glancing blow in the forehead and neatly extricated the arrow — rough surgery, to be sure, but it served. Harrington tied a rag around his head, and kept his place during the whole three days of fighting.

When they first reached the island one of the men cried out, "Don't let's stay here and be shot down like dogs! Will any man try for the opposite bank with me?" Forsyth, revolver in hand, stopped that effort by threatening to shoot any man who attempted to leave the island. In all the party there was but one coward. In looks and demeanor he was the most promising of the company—a splendid specimen of manhood apparently. To everybody's surprise, after one shot he hugged the earth in his rifle-pit and positively refused to do anything, in spite of orders, pleadings, jeers, and curses. He left the troop immediately on its arrival at Fort Wallace.

Per contra, one of the bravest, where all but one were heroes, was a little, eighteen-year old Jewish boy, who had begged to be enlisted and allowed to go along. He had been the butt of the command, yet he proved himself a very paladin of courage and efficiency when the fighting began.*

* In General Fry's entertaining story of "Army Sacrifices," the following little poem about him appears:

"When the foe charged on the breastworks
With the madness of despair,
And the bravest souls were tested,
The little Jew was there.

"When the weary dozed on duty,
Or the wounded needed care,
When another shot was called for,
The little Jew was there

"With the festering dead around them,
Shedding poison in the air,
When the crippled chieftain ordered,
The little Jew was there."

One of the last acts of the recent Congress was the setting apart of one hundred and twenty acres of land in Yuma County, Colorado, as a national park. This reservation forever preserves Forsyth's battlefield and the vicinity from settlement. On the edge of the river bank, on what was once Beecher's Island, which the shifting river has now joined to the bank, is a wooden monument to Beecher and the other scouts who were buried somewhere in those shifting sands.*

The few survivors of the battle have formed themselves into an association which holds an annual reunion on the battlefield. Soon there will be none of them left. Would it not be a graceful act for some one who honors courage, manliness, and devotion to duty to erect a more enduring monument to the memory of Beecher and his comrades than the perishable wooden shaft which now inadequately serves to call attention to their sacrifice and their valor?

NOTE.—The following interesting communication slightly modifies one of the statements in the above article. It certainly shows prompt decision upon the part of Lieutenant Johnson, who was left in command of the post after Bankhead's departure.

Great Barrington, Mass., August 5th, 1904.

Dr. Brady says "Donovan had assembled five *bold spirits*, and had immediately started out on the return journey." As a matter of fact, Donovan did no such

* "To-day cattle stand knee-deep in the Arickaree. The water no longer ripples around the island, as the shifting sands have filled the channel to the south. But if one digs under the cottonwoods he can find bullets, cartridges, and knives. And near at hand is the simple white shaft that tells where Beecher and Roman Nose, typifying all that is brave in white man and red, forgot all enmity in the last sleep that knows no dreams of racial hatred." I cut this from a newspaper the other day. How well written, frequently, are the modestly unsigned articles in the daily press!

thing. The departure of General Bankhead's relief column stripped the garrison of Fort Wallace to seven enlisted men, took away the last horse, and placed me in command. Forsyth's second note, brought by Donovan, fell into my hands. It was telegraphed in full to General Sheridan, who ordered me to spare no expense of men, money, and horses to hasten relief to Forsyth. By the promise of $100 each, four citizens of the neighboring town of Pond Creek were induced to seek the Carpenter command. Donovan I persuaded to guide them, promising him $100 in addition to his pay as a scout. This party started at daylight, on government mules, rode all day, all night, and found Carpenter's command on the south fork of the Republican River, about ten miles southeast of the scene of the fight. Guided by these men, Carpenter pushed out, and Forsyth and his men were relieved some hours in advance of the arrival of the other relief commands.

The country from Fort Wallace to Arickaree Fork I passed over the following December, in an unsuccessful endeavor to secure the bodies of those killed in the fight. We surprised a village of Indians at the scene of the fight, fought them off, and found the body of one of the scouts, but Lieutenant Beecher's and Dr. Mooers' graves were empty. Yours very truly,

HUGH M. JOHNSON,
Late Lieutenant 5th U. S. Infantry.

CHAPTER SEVEN

A Scout's Story of the Defense of Beecher's Island

BY great good fortune I am permitted to insert here a private letter to me from Mr. Sigmund Schlesinger, the Jewish boy referred to in Chapter Six, which, as it contains an original account of the defense of Beecher's Island from the standpoint of one of the participants, is an unique document in our Western historical records:—C.T.B.

For several days we had been following an Indian trail so broad that it looked like a wagon-road. Those in our command experienced in Indian warfare told us that we must be on the track of an Indian village on the move, with a large herd of horses. Evidently they knew that we were behind them, and seemed to be in a hurry to get away, for we found camp utensils, tent-poles, etc., which had been dropped and no time taken to pick them up. Among other things we saw fresh antelope meat, quarters, etc., and although our rations were nearly, if not all, gone, except some coffee and very little "sow-belly," we did not dare eat the Indians' remnants.

The night of Sept. 16th, before the attack next morning, Scout Culver, who was killed next day, pointed out to a few of us some torch-lights upon the hills that were being swung like signals. I knew that something "would be doing" soon, but, like a novice, I was as if on an anxious seat, under a strain of anticipation, expecting something strange and dangerous. The next thing that I now recall was that I was awakened just before daylight by a single cry, "Indians!" so loud and menacing that when I jumped up from the ground I was bewildered and felt as if I wanted to ward off a blow, coming from I knew not where, for it was still quite dark. That cry I will never forget. Soon I perceived a commotion among our horses and mules. The Indians, about a dozen, tried to stampede them. I could see in the dawning light the outlines of a white horse in the distance, and from the noise I realized that they were driving some of our stock before them. Later, in the daylight we could recognize some of our ponies on a neighboring hill in the possession of the Indians.

As soon as we crossed from the north bank of the river to the island, just before the attack, we tied our horses and pack mules to shrubs as best we could. During the day a mule with a partial pack on his back got loose and wandered around the vicinity of my pit. He had several arrows sticking in his body and seemed wounded otherwise, which caused him to rear and pitch to such an extent that Jim Lane, my neighbor, and I, decided to kill him. After shooting him he fell and lay between us, and served us the double purpose of food and barricade.

My horse was securely tethered to the underbrush on the island, and later that day I saw the poor beast rearing and plunging in a death struggle, having been shot and killed like the rest of our horses and mules.

He also furnished me with several meals during the siege, even after he began to putrefy. There was little to choose between horse and mule meat under such circumstances — both were abominable.

When day broke that Tuesday, the seventeenth of September, 1868, we saw our pickets riding toward camp as fast as their horses could carry them, excitedly yelling: "Indians! Indians!" As I looked up the valley toward the west I beheld the grandest, wildest sight — such as few mortals are permitted to see and live to tell about. Many hundreds of Indians in full war paraphernalia, riding their splendid war ponies, rushed toward us *en masse*. Some were galloping in one direction, others cantering in another, their lances topped with many-colored streamers, the fantastic Indian costumes lending an awful charm to the whole. About this time those among us who had any had boiled some coffee and were preparing to cross over to the island.

I will frankly admit that I was awed and scared. I felt as if I wanted to run somewhere, but every avenue of escape was blocked. Look where I might I perceived nothing but danger, which increased my agitation; so I naturally turned to Colonel Forsyth as a protector, as a young chick espying the hawk in the air flutters toward the mother wing. Under such conditions of strain some things engrave themselves vividly upon your mind, while others are entirely forgotten. I remember that distinctly as in my trepidation I instinctively kept close to the colonel. I was reassured by his remarkable self-possession and coolness. While stirring every one to activity round us, he consulted with Lieutenant Beecher and the guide, Sharp Grover, giving directions here, advice there, until most of the command had crossed; then he crossed himself and posted the men, telling them where to take up their different positions. Meantime the Indians were coming closer. I was just behind the colonel when the first shot from the enemy came flying seemingly over our heads. I heard him say, smilingly, "Thank you," but immediately afterward he ordered every one of us to lie flat upon the ground, while he, still directing, kept on his feet, walking around among us, leading his horse. The shots began coming thicker, and many of us yelled to him to lie down also. How long after this I do not know, but I heard the colonel cry out that he was shot, and I saw him clutch his leg and get down in a sitting position.

I was lying alongside of Lou McLaughlin; some tall weeds obscured my vision, so I asked Lou to crouch lower and I rolled over him to the other side and was there kept busy with my carbine, for the Indians were onto us. They were circling around while others were shooting. Very soon I heard Lou growl and mutter. I looked at him and saw that he was hit, a bullet coming from the direction where I was lying struck his gun-sight and glanced into his breast. He told me what had happened, but I could give him no attention, for there seemed lots of work to do before us. But later, after the repulse of the attack, I looked at Lou and was surprised to see him lying in a wallow. In his pain he had torn up the grass and dug his hands into the sand. In answer to my question whether he was hurt bad, he told me not bad, and advised me to dig into the sand and make a hole, as it would be a protection.

I am not sure at this time, but I am now under the impression that I told Colonel Forsyth of this; and from that time on we began to dig with our hands or whatever we could use, and kick with our heels and toes in

Courtesy of Chas. Scribner's Sons

THE CRUCIAL MOMENT ON BEECHER'S ISLAND

Drawing by R. F. Zogbaum

the sand, and some of us soon had holes dug deep enough to protect the chest, at least.

Time seemed out of our calculations. I heard some one call, "What time is it?" An answer came, "Three o'clock." I had thought it was about ten A.M. We had nothing to eat or drink all day and, strange to say, I was not hungry, which may have been the reason why I thought it was still early. Word was passed that Lieutenant Beecher and Scouts Wilson and Culver were killed, Colonel Forsyth wounded again, also Doctor Mooers shot in the head and others hurt whose names I do not now remember.

We fought steadily all day. After dark the Indians withdrew; then nature began to assert itself. I got hungry; there was nothing to eat in the camp that I knew of, except some wild plums that I had gathered the day before, which were in my saddle-bags, still on the body of my horse. I got out of my hole, creeping on hands and knees toward where I knew the poor animal lay. As I felt my way in the darkness I touched something cold, and upon examination found that it was Wilson's dead hand. He lay where he fell; it was a most horrible feeling. The shivers ran up and down my back, but I got to my horse at last, and tugging, I finally secured the bag and my plums. I also found in it a piece of bacon, the size of two fingers, which I reserved for a last emergency, and was still in possession of that rusty piece of fat when relief came.

On my way back to my hole I passed one where Doctor Mooers lay wounded, moaning piteously. I put a plum in his mouth, and I saw it between his teeth next morning. He died on the night of the 19th. All our wounded were very cheerful, and to look at Colonel Forsyth and talk to him as he lay there helpless, no outsider would have suspected that he was crippled. We used to gather round him in his pit and hold conversation, not like men in a desperate situation, but like neighbors talking over a common cause.

Colonel Forsyth was the right man in command of such a heterogeneous company. Like the least among us, he attended to his own horse when in camp, and many times have I seen him gather buffalo chips to supply the mess fuel. While he was our commander in practice he was our friend, and as such we respected him, followed and obeyed him.

On about the fifth day, as the Indians began leaving us, we began to walk about and look around. About fifteen or twenty feet from my pit I noticed a few of our men calling to the rest of us. I ran to the place, and there, against the edge of the island, I saw three dead Indians. Their friends evidently could not reach them to carry them off, which explained to us the persistent fighting in this direction. When I got there the Indians were being stripped of their equipments, scalps, etc. One of them was shot in the head and his hair was clotted with blood. I took hold of one of his braids and applied my knife to the skin above the ear to secure the scalp, but my hand coming in contact with the blood, I dropped the hair in disgust.

Old Jim Lane saw my hesitation, and taking up the braid, said to me: "My boy, does it make you sick?" Then inserting the point of the knife under the skin, he cut around, took up the other braid, and jerked the scalp from the head. I had been about three years in that country and four years in America, and life on the plains under such hardships as I had undergone hardens the sensibility, yet I was not quite ripe for such a cutting affray, even with a dead Indian.

After this we were not molested, but devoted our time to looking around for something to eat besides the rotten horse and mule meat, which we boiled several times in water and powder, not to get it soft, but to boil out the stench as much as possible. We found some cactus fruit, and killed a coyote, of which the brains and a rib were my portion. Aside from this we had nothing but horse and mule during the siege, which soon told on our bowels; but in spite of all this, I do not remember a despondent man in our crowd.

One morning, being the ninth since we were attacked, I was lying outside of my pit, having done some guard duty during the night; I was half dozing and dreaming of home and a good meal. I felt so homesick and so hungry when I heard some one call attention to something moving on the hill.

I was all attention at once. Soon I heard again "I think that's Doctor Fitzgerald's greyhound." Whoever it might be, we would welcome. We would even have been pleased to have the Indians attack us again, in hopes of killing one of their horses for fresh meat; but it was soon evident that help was coming, and when I fully realized this fact, enfeebled as I was, I jumped up and joined in a lunatics' dance that was in progress all around us. Those on the hill must have seen us, for there was a rush of horsemen down the hill toward us, followed by one or two ambulance wagons.

They were as eager to reach us as we were to greet them, and as I ran uphill I noticed a soldier on a white horse coming full tilt. The momentum carried him past me, but in passing I grabbed his saddle-bag and was taken off my feet, but it would have taken more than one horse to drag me from my hold. I suspected some eatables in there, and as soon as he could stop, without dismounting he assisted me to open that bag. With both hands I dived in, and with each hand I clutched some hardtack, but only one hand could reach my mouth; my other was in the grip of one of our men, who ravenously snatched the "tacks." We ate, cried, laughed, and ate, all in a breath.

As soon as possible we put our dead in the ground. Those that died at one end of the island were cared for by those in that vicinity, and others in their vicinity, so that one part of the island was not aware of the location of the corpses of the other part; at least I did not know where the bodies lay of those killed on the eastern end of the island. So one time, as I walked around among the pits, I noticed something red and round sticking out of the sand, like a half-buried red berry. I kicked it, but by so doing it was not dislodged; I kicked again, but to no result. I then looked closer and discovered that it was the nose of a dead man. I called others to my assistance, and we fixed matters so that no desecration was possible again.

Our mortally wounded were made as comfortable as possible before they died. I assisted at such ministrations given to Lieutenant Beecher. We removed his boots, coat, etc., and, of course, these things were not replaced on the body after he was dead, but lay around unnoticed. My shoes were quite badly worn, especially after being used for digging in the sand, so when relief came and we were preparing to leave the island, I put on his shoes, which were just about my size, and wore them even after I got back to New York City, leaving my old shoes in their stead on the island.

At one of our "sittings" around Colonel Forsyth in his pit, the incident of killing the coyote was discussed, and plans were suggested for the killing of more of them.

Along with others, I also suggested a scheme, but it was ridiculed, and I soon retired to my pit, which was near enough to the colonel's, so that I could hear what was said there. One of the men remaining was saying uncomplimentary things about me, when the colonel silenced him, telling him that I was but a boy unused to such things and that, under the circumstances, I was doing better than some of the older men. Colonel Forsyth is unconscious of the fact that I am very grateful to him for his kindness to that strange "boy" among those strangers, and that I still hope some day that I may have the opportunity to show my appreciation.

Jack Stillwell and I were the only boys in the company, and naturally gravitated toward each other. We were friends as soon as we met and chums before we knew each other's names. When the colonel asked for volunteers to go to Fort Wallace for help, Jack was among the first to announce himself. I wanted to go with him, but the colonel gave no heed to my request; even Jack discouraged me, for he knew I was too inexperienced. After Colonel Carpenter came to our relief Jack was not with him, which made me and others feel very uneasy. The day after Colonel Carpenter's arrival we saw the mounted sentinel that had been posted by Colonel Carpenter on a high eminence in the hills about three miles from the island, signaling that a body of men was approaching, which created a flutter of excitement, but there was a strong sensation of security, mingled with a sense of dependence upon our black rescuers permeating our emaciated party, after being cooped up, so to say, for so long a period in dread and suspense. At least that was my sensation. I remember watching that vedette, horse and rider turning around and around, being the only moving object in that dim distance, indicating to the anxious watchers that either friend or foe was in the vicinity. As he showed no inclination to leave his post, it was soon evident he had no fear of the approaching column, and that friends were coming. Not long after a few horsemen were seen coming around the bend of the river bed, and among them was my friend Jack Stillwell. Nearly all of us ran to meet the party. Soon Jack jumped from his horse, and in his joy to see so many of us alive again, he permitted his tears free flow down his good honest cheeks. I kept up correspondence with him all these years past. Last year he died.* He was a big-hearted, jovial fellow, brave to a fault.

* Stillwell studied law, and ultimately became a judge in Texas. He was a friend of Generals Miles and Custer — also of "Wild Bill" Hickock, "Texas Jack" Omohundro, and other famous figures on the frontier; and when he died, a couple of years ago, he was the subject of glowing tributes from high and low alike.—C. T. B.

CHAPTER EIGHT

Carpenter and His "Brunettes."* The Fight on Beaver Creek

CARPENTER had performed a very commendable thing in his march of over one hundred miles in two days for the relief of Forsyth. And it is marvelous that he had been able to find him in that vast expanse of country. He received high praise for it, which he fully deserved; but the battle which, with his black comrades, he fought three weeks later, elicited still more praise. The fight was one of the prettiest and most typical of any in our Indian campaigns; and I am fortunate in being able to give it in General Carpenter's own words, written especially for this book, the notes appended being my own contribution.—C. T. B.

While on the forced march to relieve the party of scouts with Colonel George A. Forsyth, surrounded by Indians on the Arickaree fork of the Republican River, the troops under my command discovered a large trail of the Indians who had been engaged in that fight on the south fork of the Republican.

*Negro troops were often so styled by their white comrades in the service.—C. T. B.

The scouts discovered that this trail left the valley of the stream a short distance below and struck across country in the direction of the Beaver Creek. After the relief of Forsyth, on my return to Fort Wallace with the survivors and wounded, a report was made to General Sheridan, then to the east of Fort Hayes, Kansas, of the probable whereabouts of the Indians; and the Fifth Cavalry, which had just arrived from the East by rail, was disembarked between Hayes and Wallace and ordered to move north under Major Royall, and strike the savages, if possible, on the Beaver. A day or two after the Fifth had left, Brevet-Major-General Eugene A. Carr reported for duty to General Sheridan. Carr had been a general officer of volunteers during the Civil War with an excellent record, and now reverted to his rank in the regulars of Major in the Fifth Cavalry.

Sheridan was anxious to have Carr join his regiment because of his experience with Indians and his general reputation, and therefore sent him to Fort Wallace with orders to have the two troops of cavalry there go under my command and escort Carr and overtake the Fifth, if possible, to enable him to join his regiment. The troops consisting of Troops H and I, Tenth Cavalry, were officered by myself and Captain Graham, Lieutenants Banzhaf, Amick, and Orleman, and were soon in readiness for the duty required. We had returned from the relief of Forsyth Oct. 1st, and we started with Carr at ten A.M. Oct. 14th.

I concluded to march north so as to strike the Beaver as soon as possible, and then to follow down that creek with the expectation of finding the Fifth Cavalry or of striking its trail. On the 15th I reached the Beaver at about one P.M., and after proceeding some miles down, went into bivouac. As we expected, we found a very large Indian trail about two weeks old, over which over two thousand head of ponies had been ridden or driven, going in the same direction.

The next day we continued our journey down the stream, finding plenty of water, a fine bottom covered with grass and timber, and still observing the Indian trail, which ran to a point about twenty miles east from the place where we first struck the Beaver. At this locality the signs showed that the Indians had encamped for the night. The ground was covered for acres with old fire-places, pieces of wood, and the manure of ponies; and a little distance off we found a dead Cheyenne, wrapped in his robes, lying upon a scaffolding in a tree, a protection against ravenous wolves. The trail then struck south toward Short Nose Creek, the Indian name for a stream about twenty miles south of the Beaver. We continued our course, however, on the Beaver, until we made about thirty miles, and then stopped for the night.

As there was no pack outfit at Wallace, I was compelled to take wagons to carry our supplies, and had eleven with me. The mules, dragging heavy loads over rough country, were made to trot in order to keep up with the cavalry column. We had now moved down the Beaver about forty-five miles without finding anything about the Fifth, and it began to look as if something had taken the regiment in another direction.

The next morning I sent Lieutenant Amick and ten men well mounted with Sharp Grover, the famous scout, with orders to proceed as quickly as possible across country to the Short Nose to look for signs of the Fifth Cavalry and to keep a sharp lookout for Indians.

Grover, who, it will be remembered, had been with Forsyth, afterward joined my command. He had married a Sioux woman and had lived for years with the Indians before the outbreak of hostilities. He could speak their language and knew their ways and customs, and was perfectly trained in reading signs. It was interesting to see how he could read what the tracks meant, as if they had been books. He could tell how long since the tracks were made, whether they were made by horses or ponies, shod or unshod, how many were ridden, how many were driven, whether it was a war party or a party changing camp. If Indians stopped for the night he could tell how many men or squaws were in the party, to what tribe they belonged, from the shape of their moccasins, and many more details. Like most of his ilk, Grover drank heavily on occasion. When the Indians went on the war-path Grover could not stay longer with the Sioux, as his life was not safe, and he entered the government employ, where he rendered heroic and invaluable services. Later he was killed in a row at Pond City, near Fort Wallace.

Amick and his party soon disappeared over the hills to our right and we kept on down the stream, the general course of which was to the northeast. I began to feel certain that the Fifth Cavalry had never reached the Beaver, and that we would probably be attacked by the Indians if this was the case. Under these circumstances I felt that it would be wise to be cautious and on the lookout against surprise. The road we passed over was very rough and the stream in most places ran through deep-cut banks several feet high, with very few places suitable for crossing.

As night came on a place was selected for a camp in a bend of the creek where the wagons could be placed across, giving room inside to graze the animals without fear of a stampede from howling savages. Amick returned just before night, having scouted some miles beyond the Short Nose without discovering any trace of the Fifth Cavalry. Grover told me that as they passed across the divide between the Beaver and the Short Nose he came across a single Indian pony track. This track was coming from a direction to our rear, and showed that the pony was going at a rapid gallop. Grover inferred from this that it was probably an Indian hunter returning homeward who had most likely crossed our trail behind us, discovered our presence in the country, and was riding as fast as possible to carry the news to the Indian camps somewhere to our front and not far off.

After a council over the situation General Carr came to the conclusion, after having traveled some sixty miles down the Beaver without finding the Fifth, that the regiment had never reached that stream and that therefore he would give it up and start on our return in the morning. About seven A.M. on the next day, Oct. 15th, Captain Graham expressed a wish to make a scout for a short distance to the front, and rode forward with two men. The command was ordered saddled up and everything made in readiness to move. In view of the fact that the south side of the creek was hilly and difficult and offered opportunities for ambuscades, I determined to go back by the north side, which was comparatively open. The afternoon before I had sent Lieutenant Orleman with a detachment to dig down the sides of the creek and prepare a practicable passage for the wagons and troops.

Graham had hardly ridden a thousand yards when twenty-five Indians suddenly dashed over the hill to his rear, with the evident intention of cutting him off. They were almost upon Graham before he discerned them, but he instantly struck spurs into his horse and dashed for the creek, the Indians firing a volley at short range upon the party. One of the bullets passed through Graham's hat, another through his coat, and a third through his leggings without wounding him. One of the horses was shot through the shoulder and fell. His rider succeeded in getting into the creek and behind the bank along with the other soldiers, and they commenced firing upon the Indians. Graham's girth burst as his horse sprang away at the first fire, but as his saddle gave way he seized his horse's mane and dragged himself forward on the animal. He then dashed the horse over the bank of the creek, about ten feet to the bottom. He fell from his horse in this jump, but the horse, fortunately, ran in our direction.*

By this time I started out thirty men under Lieutenants Amick and Orleman to cover the retreat of Graham's men. As they charged toward the hill the savages rushed from the creek to avoid being cut off, and were hotly pursued by our men. Judging that the presence of these Indians indicated that a large party could not be far off, I thought it best to be prudent and sent a trumpeter to overtake Amick and tell him to discontinue the pursuit and fall back slowly to camp. Without further delay I now broke up the camp, crossed the creek with wagons and troops, and, having dismounted the men, deployed them as riflemen to cover the retreat of Amick.

In a few minutes the absent party made its appearance on the hills, with bodies of Indians, numbering at least a hundred, skirmishing on our flank and rear.

* Luckily for Graham, just as he fell from his horse Amick's men charged the Indians, who thereupon retreated. One of the troopers caught the horse and held him until the captain came running up. Carpenter's prompt action in instantly throwing Amick's detachment on the savages undoubtedly saved the lives of Graham and his men. As it was, it was a fearfully narrow escape for them.— C. T. B.

They slowly fell back toward the creek, and when within range the dismounted men on the banks opened fire on the advancing savages, and under cover of this Amick crossed and joined the command, while the Indians kept at a respectful distance.

The wagons were now placed in double column so as to make everything as compact as possible.* H Troop was assigned on the flanks and advance, deployed in open order. Troop I covered the rear in the same manner, with one platoon under Graham as a reserve. These arrangements being completed, we moved steadily up the creek bottom. As soon as this movement commenced, a large body of Indians made their appearance and charged toward us, taking advantage of ravines, trees, and bluffs to fire from the south side of the creek. Some of the balls were well aimed and came close.

I soon saw that if we continued down the creek bottom the enemy would harass us immensely under cover of the timber and banks, and therefore changed our course so as to leave the valley and take the higher ground or divide. The Indians followed, showing about two hundred strong, and acted boldly in their attacks on the rear and flanks. The men and officers behaved very coolly, facing toward the enemy and driving them back without stopping the progress of the column.

At one point we passed near a deep ravine, and the enemy, quick to observe cover of any kind, occupied it with quite a number of warriors and opened up a serious fire. The reserve platoon under Graham charged at the place as we were passing and, arriving at the edge of the ravine, poured in a volley at close range on the

* Six wagons on one side, five on the other.—L.H.C.

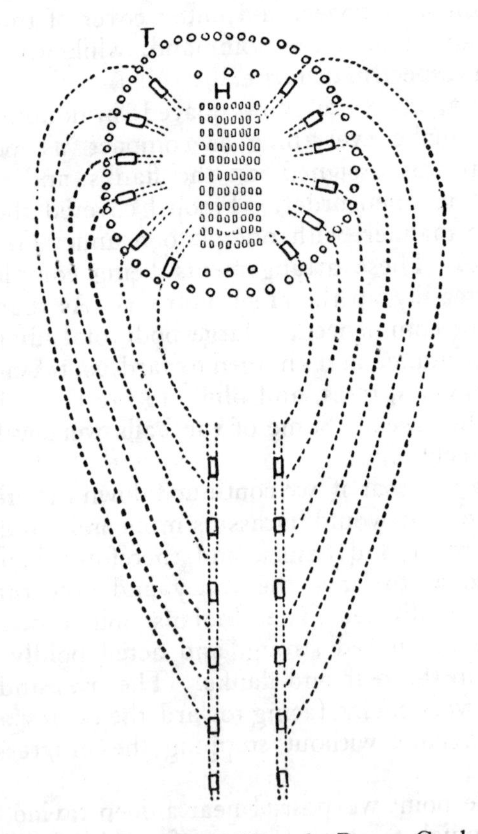

Position of Wagons and Soldiers in Beaver Creek Fight

T, Troopers Surrounding Wagons; H, Horses Inside the Inclosure

Drawn by General Carpenter

savages. A number must have been hurt and the Indians certainly lost no time in getting out of their position. Afterward they were more cautious in occupying ground too close to us. The flankers, under Banzhaf and Orleman, also repulsed the Indians on several occasions.

One Indian carried a red flag with some white device upon it, and by his movements the whole force seemed more or less governed. They were all stripped to the waists, and were decorated by various ornaments hanging from their heads and their shields, quivers, and bridles, so as to glisten and shine in the sun at every turn of the ponies. Up to this time five Indians were known to be killed at various points and quite a number wounded.

At one P.M. the enemy seemed to stop the fight and apparently withdrew, and I supposed that I had seen the last of them; but half an hour afterward, hearing an exclamation, I looked back and saw the Indians appearing again on the hills to our rear. On they came, one body after another coming in sight until it was estimated by all present that at least six hundred warriors were in view. Emboldened by their number they rushed forward, directing themselves toward our front, flanks and rear, making things look rather serious. I soon saw that we could not continue the march and meet this force, but that we must select a position and make a stand.

In the first attack in the morning I had offered the command to General Carr, as the senior officer present, but he declined it, stating that he considered himself simply as a passenger to be escorted, and I therefore continued to direct the operations.*

* General Carr, while he used a rifle efficiently during the whole of the action, did not interfere with Carpenter's arrangements; with rare self-restraint he refrained even from

I looked around and saw a small knoll or rise a short distance to the front, from which the ground fell in every direction, and this point was immediately selected. The teamsters were directed to take the trot, aim for this place, and on arrival at the knoll immediately to form a circular corral, half a circle on either side, with the mules facing inward, affording a shelter within and something of a fortification. As soon as we increased our pace the Indians evidently thought we were running from them, and sent up a yell which made shivers run down the backs of some of our recruits. We kept on, however, at a fast gait, and the moment we struck the highest ground, the wagons were corralled with six wagons one side, five wagons on the other, and the troops were rushed inside at a gallop and dismounted. The horses were tied together inside the corral with some men to watch them, and the rest were formed outside the corral in open order.* This was done in about two minutes and then the advance of the Indians was upon us.

A fire commenced from our seven-shooter Spencers which sounded like the fire of a line of infantry. The

offering suggestions. Although he showed plenty of nerve and was entirely calm and collected throughout the hot affair, he is reported to have said after the fight that he thought he was in the tightest box in which he was ever caught in his life, and that there was nothing left for them all to do but to sell their lives as dearly as possible. He had never served on the plains with negro troopers before, and had no knowledge as to how they would behave against Indians — nobody had, for that matter. — C. T. B.

* When there are many wagons, the corral is formed with the wagons close together, axles touching, tongues and mules inward; but in this case they had to form an open corral with about twenty feet between the wagons. The horses, to the number of one hundred — the troops being reduced to that number by illness, hard work, and other causes — were forced inside the wagons in columns of four, each set of fours being tied together, with horse-holders detailed to look after them. The duty of the horse-holders was most important. They had always to be on watch to prevent a stampede. If the troops had been deprived of their horses on that open knoll their position would have been hopeless indeed. The dismounted men were formed in front and rear and on both sides of the corral, as Indian attacks must be expected from every direction. The plan made by General Carpenter explains the situation perfectly. — C. T. B.

THE CHIEF MEDICINE MAN AT BEAVER CREEK

Drawing by Will Crawford

Indians charged up around the wagons, firing rapidly and seriously wounded some of the men, but in a very short time they were driven back in wild disorder, leaving the ground covered with ponies, arms, and some bodies. Three dead warriors lay within fifty feet of the wagons. One man who was killed here was carried off by his comrades.

The chief Medicine Man, on a fine looking horse, rode out in front of our line about two hundred yards off, after the retreat of the Indians, to try to show that his medicine was good and the white man's bullets could not hurt him. I directed several men near me to aim carefully at him. They fired and the Medicine Man went down, accompanied by a howl from the more distant Indians. After the repulse the men rushed forward from the wagons, seized and hauled in ten bodies of the Indians. The savages, disheartened and surprised at this reception, withdrew out of gunshot and assembled, apparently for council.*

The men carried corn sacks and made breastworks near the wagons and we waited, expecting a renewal of the attack, for about an hour, when it became evident that some of the Indians were withdrawing. The day was very warm, we had been engaged about eight hours, and in the hot sun men and animals were suffering very

*Just after the first repulse of the Indians, Graham went to General Carr very improperly (he had no business to communicate with him except through Carpenter, his immediate commanding officer), and suggested that he order the wagons abandoned, the teamsters mounted on the mules, and the whole party to retreat in that fashion as fast as possible. Carr, of course, referred the matter to Carpenter, who promptly refused to sanction any such manœuver. To leave the wagons, which afforded some protection on the hill, would be to enable six or seven hundred Indians to surround the hundred troopers in his command and invite annihilation. Carpenter assured Carr that as he had defeated the Indians once, he was certain that he could do it again. Carr wisely persisted in his determination not to interfere, and Graham had his suggestion decisively negatived by Carpenter. Graham deserved reprobation for his unmilitary action, *as well as for his foolish suggestion.*— C. T. B.

much from thirst. I made up my mind to move for water, and keeping the wagons in double column, the horses inside and the men dismounted on the outside, we marched for the Beaver. A large party of Indians followed up to where their dead comrades lay and set up a mournful howl over their remains. Their loss in this fight, added to what they had suffered the month before in the conflict with Forsyth, must have had a sobering effect.

We now proceeded to the creek without further interference, and selecting a wide bottom encamped for the night, preparing some rifle-pits to cover our outlying pickets and to enable them to receive the enemy if an attack were made in the morning. We heard them around us all night imitating coyotes, but they did not find a weak place and refrained from molesting us. The next morning the Indians were gone and we marched by the shortest route to Fort Wallace, arriving there on the 21st.

On our return journey we passed through Sheridan City, a frontier town located at the then terminus of the Kansas Division of the Union Pacific R. R. It was full of taverns, saloons, gambling houses and dens, and of a rather tough lot of citizens and desperadoes. These people and others crowded into the streets when we passed through, and when they saw the troopers and their horses decorated with the spoils from the Indians whose dead bodies we had captured, they knew that we had been in a successful fight and they gave us a perfect ovation.

The savages suffered a considerable loss, but we escaped with a few men wounded (some of them seriously) and none killed. General Carr found the Fifth Cavalry had returned to the railroad, and through mistake they never reached the Beaver. He took command of the regiment, marched again and pursued the Indians over the Platte River, and followed them on a long campaign.

This was one of the smartest and most successful Indian fights on record. Carpenter's tactics throughout had been admirable. General Carr was much surprised and pleased at the conduct of these black troopers, and on his return to Fort Wallace telegraphed to General Sheridan that "the officers and troops behaved admirably." General Sheridan published a general order highly commending the commander, the officers and the men for this brilliant and gallant affair. Carpenter was brevetted colonel in the Regular Army (his fifth brevet), and afterward received a medal of honor for this fight and the relief of Forsyth. Well did he deserve them both.—C. T. B.

CHAPTER NINE

A Further Discussion of the Beaver Creek Affair

WHEN General Carpenter's account of the fight on the Beaver Creek was published serially, General Carr took exception to it in a public letter to the editor of the periodical in which it appeared. I am permitted by the editor to make extracts from this letter, which, with my own comment and General Carpenter's rejoinder, appear as follows:

I. General Carr's Account

I do not think that General Sheridan sent orders for my escort to go under command of Colonel Carpenter. I know that, after waiting with General Sheridan at Fort Hayes for several days, he ordered me to go to Fort Wallace, take an escort and go to find Royall, who had not been heard from as expected.

While we were going over the "rough hills" Colonel Carpenter, and perhaps some of the others, came up and remonstrated with me for marching the command so hard; said they could not be responsible for its effect on the horses, and when I insisted, asked me to take command and be responsible. I said, "All right, I will take command; but you must attend to the details. I will not appoint an adjutant and take the office work" (detailing guards, stable duties, etc., etc.).

I had with me an officer of my regiment, Captain Kane, Fifth Cavalry, and now wish I had appointed him adjutant and taken formal command.

Finding that the stream (Beaver Creek) was persistently hugging the south side of the valley, on which we were traveling, creating bluffs and ravines over which it was difficult to move the wagons, I determined to cross to the north side, where the valley consisted of a gentle and smooth slope.

I selected the camp in an "ox bow" bend of the creek, putting the tents and wagons across the entrance, the guard at the bow and the animals inside, selected a place for crossing, and ordered the banks to be cut down to the bed of the creek, so that the wagons and animals could cross.

In the Indian country I always had my commands up and under arms before daylight, which is the most dangerous time; and next morning we were up, had breakfast, the tents struck and the wagons packed, and were standing 'round the camp fire when Amick called out "Indians!"

I had intended to move camp across the creek and get fresh grass. Graham had started on his trip, following a narrow path between the bluff and the river bank. He was riding a fine horse called "Red Eye," which had a very deep chest and thin barrel. As he told me afterward, his saddle began to slip back, but he waited to get down to fix it till he should reach a more open place ahead at the mouth of a ravine, when *bang, bang!* came some shots. His horse dodged and jumped down the bank into the creek, slipping the saddle farther back and kicking it clear, while Graham landed on the shoulders of the horse and from there on the bed of the creek. He turned back with his gun toward the bank, which protected him while he fired, till I sent Lieutenant Amick with some skirmishers, who covered his retreat. His two men jumped their horses down into and up out of the creek and fell back, and when we got the men all in I moved the command up to and over the crossing I had prepared and out to the open slope on the north side of the valley. As we had not found Royall but had found the Indians, or, rather, they had found us, it was of no use to go farther down the Beaver, and I determined to move toward home.

The Indians kept coming out from those bluffs, crossing the creek and following us, and I arranged the wagons in two columns, with Carpenter's company in front and on the flanks, and Graham's company in rear, where I was also.

Some of the Indians got on our flanks as skirmishers, a few in front and more in rear, while the main body got into two columns, about like the two halves of a regiment in columns of four, marching steadily and gradually, closing on us by increasing their gait. I estimated them at seven hundred warriors; we had about one hundred colored soldiers. Their chiefs were marching between the heads of their columns; they had a flag and a bugle. I sent word to Carpenter to turn more toward the higher ground, away from the timber along the creek, which Indians were using for cover. Pretty soon he began to trot, and I had to send Captain Kane to him with orders to go slower, for fear of a stampede. I was in the rear near Graham's troop. He was one of the bravest men I ever saw. He would amuse himself with the Indians by concealing men in hollows, taking away their horses till Indians came up for them to shoot at, then charging up with their horses, mounting and bringing them off. Most of the men were cool, but I saw one man loading and firing when the Indians were a mile away. I called to him to stop, but he paid no attention, seemed to be dazed, and I had to go up and lay my hand on him before he obeyed my order. As the Indians got closer, one of the men dismounted to shoot. When he tried to remount, his horse got scared and edged away from him, and I rode up on the off-side, got hold of the rein, and let his horse come against mine. Then he put his foot in the stirrup, made a leap and got his leg over, when we both cantered off — just in time. He lost his cap, and we lost a dog which was playing about between us and the Indians, interested in the shooting till he got near enough to them to be killed. These were our only losses.

As the main body of the Indians got closer and increased their gait to the charge, the leading wagons turned toward each other and stopped. The others turned their mules inside of those in front of them, lapping onto and tying to the wagons, forming a corral in shape of a flat-iron, and the horsemen rode inside, dismounted, tied their horses, and began to fire at the Indians who were thronging around us. One Indian, on a clay-bank pony and wearing a red blanket, with no arms, rode 'round and 'round us within fifty or one hundred yards, and seemed to bear a charmed life. I shot at him several times, and Grover said he also shot at him till he began to think the Indian had strong medicine which protected him — Grover was a squaw man and had imbibed some of their superstitions. He was shooting a Sharp's carbine, and, taking it down to reload, happened to notice that he had the sight set for five hundred yards and had been shooting over the Indian's head.

I had a Spencer carbine which had been sent me by the company for trial. With it I shot down an Indian, who fell and lay within thirty yards of the corral. He was not dead, and I afterward talked with him through Grover. He said his name was Little Crow, and mentioned the name of his father, whom Grover said he knew. He was young — about eighteen. I asked him why they attacked us. He said because we came on their creek. He told where their camp was, so that one week from that day, which was Sunday, October 18th, 1868, I found them with my own regiment, the Fifth, had a fight lasting two days and two nights, and drove them out of Kansas.

While the fight was going on our soldiers showed great bravery. While inside the wagon corral they would rush from one side to the other, wherever the Indians appeared, so as to fire at them. Their officers did not seem to keep them in their places, and, after it was over, I reprimanded them for not commanding their men. After the Indians were repulsed we went around and rearranged the wagons, got out sacks, cracker-boxes, etc., fixed breastworks in anticipation of another attack, and then sat down and ate our lunch.

It was two o'clock. Our soldiers wanted to scalp the dead Indians, of whom about ten were lying too near the corral for their friends to carry off. We prevented this till, while we were going around the corral rearranging, the soldiers got the chance to scalp them. Two were lying wounded when two soldiers approached them; one drew his bow and sent an arrow through the thighs of one of the men. The arrow passed through the fleshy part of one thigh and entered the other and stopped against the bone. It had to be cut in two to extract it. Years after, at Fort Leavenworth, a first sergeant showed me an arrow-head with the point marred, and said he was that man.

The other wounded Indian had a pistol, and shot the other soldier in the calf of his leg. These were the only men we had hurt. It was then that there was some talk of abandoning the wagons and going direct for the railroad. With the wagons, we had, of course, to go by routes over which wheeled vehicles could travel. I do not recall whether Captain Graham advocated this, but there would have been no impropriety in his stating his views to me. Any officer can talk to his commander, and I had been giving orders from time to time directly to him and his men. We sat there eating our lunch and talking with each other and the wounded Indian who lay in front of us, when he made a motion with his hand back toward me. I asked Grover what he meant. He said, "He wants you to go away; says his heart is bad." This pleased me very much. I had been watching the Indians, who had fallen back to some rising ground nearly a mile off. They were moving about, pow-wowing, and, I supposed, preparing for a new attack. They seemed to be diminishing in numbers, but I thought they were trying to get around to some place where they would have a better chance at us. I really did not expect to get out of that fix. If those Indians had had sufficient resolution, being seven to one, they ought to have used us up. When the wounded Indian made this motion, I took it to mean that he knew they were giving up. He could see them as well as we could, and knew better what they were doing. I suppose that those who remained were the friends of the dead, waiting for us to go.

I gave orders to reload the wagons and move out. Some of the Indians followed us for several miles, but did not again attack.

Colonel Carpenter did not offer the command to me at the first attack in the morning, nor did I decline it. I was exercising the command all the time. One of the articles of war provides that "when troops join and do duty together the highest in rank of the line of the army shall command the whole, and give the orders needful for the service," and I could not have avoided the responsibility.

The foregoing narrative shows that I was not a "passive spectator," nor did I "refrain from advice or suggestion," nor from giving orders as required.

I, no doubt, said that I was a guest, and did not interfere unnecessarily; but my long experience on the plains and with Indians rendered it incumbent on me to exercise my judgment.

EUGÈNE A. CARR,
Brevet Major-General and Congressional Medallist.

With reference to this letter from General Carr I beg to point out:

That General Carr in his letter practically admits in two places that he was not in command. First, when he writes of Captain Kane, "I wish I had appointed him adjutant and taken formal command." Second, when he writes: "I, no doubt, said that I was a guest, and did not interfere unnecessarily."

I do not see, therefore, that there is really any serious discrepancy between the account of General Carpenter and that of General Carr. I presume, if there is, I am more to blame than any one for the note to which General Carr takes exception. The original wording of that note was, possibly, not happy, and probably conveyed more than I intended. I did not suppose that any one thought that General Carr sat around and twiddled his thumbs while the fighting was going on. I have made changes in the note, which appears in this book in its amended form.

It must not be forgotten that General Carpenter received his medal of honor in part for this fight, and he certainly would not have received it had he not been in command.

Here follows a communication from General Carpenter on the subject. This discussion, I think, settles the matter in a way which I trust will be satisfactory to the friends of both of these distinguished officers.

II. General Carpenter's Reply

I was very much surprised in reading General Carr's contention that he was in command of the troops who constituted his escort to enable him to join the Fifth Cavalry, and who were in the action on the Beaver in October, 1868. The account I wrote is from my recollection, and from letters written home at the time, and I repeated what I thought there was no question about, and had no desire to ignore General Carr or any one concerned.

The two troops of the Tenth Cavalry, a portion of the garrison at Fort Wallace, under my command as the senior officer, were ordered to act as escort for General Carr, then Major, Fifth Cavalry.

I am quite positive that General Carr was offered the command by me, knowing him to be senior, and that he declined it. Many officers have been escorted by troops without taking command of the escort although of senior rank. As the troops were to escort this officer, of course he was consulted about the direction of the march, the time for camping, and concerning many other details of greater or less importance, but he never took formal command. If he did not, then, it would be manifestly improper for him to issue orders directly to subordinate officers or men. The officers and men were mentioned in General Orders by General Sheridan in 1868 for their gallantry and bravery in this action, and it is stated that they were escorting General Carr.

Thirty-five years have passed since the affair, and I never heard of any question about who commanded the troops engaged. This length of time naturally accounts for some difference of memory, and no two narratives may be expected altogether to agree. As mentioned before, Carr was consulted frequently during the expedition and his wishes were carried out, as was proper under the circumstances.

Some of the details mentioned by General Carr I do not remember, but I certainly regarded myself as being in command of the troops during the fight and gave directions as I have stated in accordance with my recollection, and selected the ground on which the wagons were corralled and the Indians repulsed.

I do not know of any officers being reprimanded, but when we returned to Fort Wallace we were informed that General Carr had reported to General Sheridan that "the officers and men had behaved admirably."

I inclose a copy of General Fleld Orders, No. 4, Headquarters Department of the Missouri, dated October 27, 1868, issued by order of General Sheridan to commend the conduct of the troops engaged in the combat on the Beaver, October 18, 1868:

Headquarters Department of the Missouri.
In the Field, Fort Hays, Kansas, October 27, 1868.

General Field Orders, No. 4.

The attention of the officers and soldiers of this department is called to the engagement with hostile Indians on Beaver Creek, Kansas, October 18, 1868, in which a detachment of cavalry (escorting Brevet Major-Gen'l E. A. Carr, Major, Fifth Cavalry to his Regiment) under the command of Brevet Lient-Col. Louis H. Carpenter, Captain, Tenth Cavalry, consisting of Companies I, Tenth Cavalry, under Capt. George W. Graham and 1st Lieut.-Major J. Amick, and H, Tenth Cavalry, under 1st Lieut. Charles Banzhaf and 2d Lieut. Louis H. Orleman, engaged about five hundred (500) Indians for several hours, inflicting a loss on the savages of ten (10) killed and many wounded, losing three (3) enlisted men wounded.

The major-general commanding desires to tender his thanks for the gallantry and bravery displayed by this small command against so large a body of Indians.

By command of Major-General Sheridan.

J. Schuyler Crosby,
Brevet Lieut.-Col. A. D. C.
A. A. A. General.

It is expressly stated that the detachment of cavalry (escorting Brevet Major-General Eugene A. Carr, Major, Fifth Cavalry, to his regiment) was "*under the command of Brevet Lieut.-Col. Louis H. Carpenter, Captain, Tenth Cavalry.*"

The records show, further, that Brevet Lieut.-Col. Louis H. Carpenter was brevetted colonel for gallant and meritorious conduct in the engagement with Indians on Beaver Creek, Kansas, October 18, 1868. This appointment was made on the recommendation of General Sheridan, and was undoubtedly conferred on this officer as being in command of the troops during the fight. Others behaved gallantly and their conduct deserved recognition, but this was the only brevet given at the time. Louis H. Carpenter.

CHAPTER TEN

The Battle of the Washita

I. Custer and the Famous Seventh Cavalry

A FIGHTER of fighters and a soldier of soldiers was that *beau sabreur* of the American Army, George Armstrong Custer, "Old Curly" to his men, "The White Chief with the Yellow Hair," or, more briefly, "Long Hair" to the Indians. From Bull Run to Appomattox his career was fairly meteoric. Second lieutenant in the Army of the Potomac at twenty-one, fresh from West Point, a brigadier-general at twenty-three, a major-general at twenty-four, and commander of the third cavalry division, which, in the six months preceding the downfall of the Confederacy, had taken one hundred and eleven guns, sixty-five battle-flags, and over ten thousand prisoners of war, without losing a flag or gun, and without a failure to capture whatever it went for — such was his record.*

I have heard my father tell of the impression made by the dashing young soldier whose spirited horse ran away on Pennsylvania Avenue at the Grand Review in Wash-

* This statement has been called in question. The facts are taken from Custer's farewell order to his division, April 19, 1865, as published in Captain Frederick Whittaker's "Complete Life of General George A. Custer," Sheldon & Co., New York, 1876. There is no possible doubt as to the correctness of the statement.

ington, in spite of the efforts of his rider — a peerless horseman — to restrain him. Custer's hat fell off, his long, yellow curls floated back in the wind, making a dashing and romantic picture. He was a man of superb physique and magnificent strength. I saw him when I was a boy, and I have never forgotten him. His devoted wife, in one of the three charming books in which she has told the deathless romance of their married life on the frontier, relates how, on one occasion, riding by her side, with his left arm he lifted her out of the saddle high in the air, held her there for a moment or two, then gently replaced her on her horse. No fatigue was too great for him to surmount, no duty, however arduous, ever caused him to give back.*

Reams have been written about his unfortunate campaign upon the Little Big Horn, in which he went down to such awful destruction, but little is known of some of the exploits of his early career on the plains. After the war, more fortunate than most of the younger general officers who were forced to content themselves with captaincies or less, General Custer was appointed Lieutenant-Colonel of the new Seventh Regular Cavalry, a regiment which was born with him, lived with him, and a large part of which died with him.

The officers of the regiment were a set of unusual men. Custer himself was allowed considerable voice in the selection of them, and such a body of officers had been rarely assembled in one command. Most of the troopers were not at first of the high grade to which they afterward attained. The best men, in the ranks at least, at the close of the Civil War, had had enough of

* It is interesting, in view of his great services to his country, to learn that the first American ancestor of the Custer family was a Hessian officer who was captured at Saratoga in 1777.

fighting. They wanted to get back to civil life once more. Not frequently it was only the inferior soldiers who could be induced to re-enlist from the volunteer into the regular regiments which were being organized or reorganized.

There were in the ranks, however, a leaven of veterans who were soldiers from love as well as from habit. With these as a nucleus, Custer and his officers, by a judicious weeding out and a rigorous course of discipline, soon gathered a body of troopers than which there were none finer in the service of the United States, nor, in fact, in any other service. Owing to the fact that the colonel, a distinguished general officer in the war, was on detached service commanding a department, the regiment was practically continuously under the command of Custer until his death in 1876.

The duty that devolved upon it was the protection of the settlers in Kansas. The job was no sinecure. In the last half of the year 1868 statistics, which do not pretend to be comprehensive, for they are only facts reported officially to the headquarters of the Department of the Missouri, show one hundred and fifty-seven people killed, fifty-seven wounded, including forty-one scalped, fourteen women outraged and murdered, one man, four women and twenty-four children taken into captivity, one thousand six hundred and twenty-seven horses, mules and cattle stolen, twenty-four ranches or settlements destroyed, eleven stage coaches attacked, and four wagon trains annihilated. This with a total loss to the Indians of eleven killed and one wounded. Truly there was a reign of blood upon that frontier. Every man murdered was also frightfully and disgustingly mutilated. This record takes no account of soldiers who were killed.

In one instance ten troopers under Lieutenant Kidder, of the Second Cavalry, with a message for Custer's command, then in the field, were overtaken and slaughtered to a man after a desperate defense. When Custer came upon the scene of battle the bodies were so mutilated that it was impossible to tell one from the other. The only distinguishing mark upon any one of them was a shirt neckband made of a material of a peculiar marking, which was yet a common article of wearing apparel at that time. It was by this shirt collar that the body of Lieutenant Kidder was subsequently identified by his mother and taken East for burial.

As usual, there was strife between the Indian agents and the army. There always has been, there always will be. The agents invariably declared that there was peace in the land and sought to embarrass the army in its efforts to protect the frontier. Popular indignation, however, at last forced the government to act, and the campaign was long and arduous during the latter part of the summer of 1868.

The success of the soldiers was not pronounced at first. The extent of territory was great, the force available small, the Indians exceedingly mobile, and the troopers had as yet scarcely learned the rules of the game, so that it was a matter of extreme difficulty to get at any considerable body of Indians and inflict a crushing blow. As we have seen, General Forsyth's command barely escaped annihilation in the great battle of the Arickaree. Matters dragged on, however, with nothing decisive happening until the summer and fall had slipped away and winter was at hand. The Indians rarely did any fighting in the winter. It was difficult and dangerous for horsemen to move on the exposed prairies in the winter season, and hitherto fighting had been abandoned with the advent of the cold. The Indians, during the winter, naturally tended southward, seeking a less severe climate if it might be had, and from November to April had been considered a closed season.

II. The March in the Blizzard

General Sheridan, however, who had command of the department, determined to inaugurate a winter campaign in the hope that the Indians, who would naturally congregate in large villages in secluded spots sheltered by trees along the river banks, might be rounded up and defeated decisively. The force at his disposal for these projected operations consisted of eleven troops of the Seventh Cavalry, four companies of infantry, and the Nineteenth Kansas Volunteer Cavalry, a regiment of settlers and old soldiers which had been organized for the campaign.

The expedition was under command of Sheridan himself. The rendezvous was at Camp Supply, in the Indian Territory, about one hundred miles south of Fort Dodge, Kansas. No Indians—in any considerable body, that is—had been seen by any of the scouts sent out, and no outrages were reported. It was evident that the hostiles were lying snugly concealed somewhere for the winter season. Sheridan determined to detach Custer and his regiment from the command and send them scouting farther southward, while with the rest of the force, so soon as it should be in condition to march, he himself would explore the country in other directions.

Custer received his orders on the 22d of November, late at night. Reveillé was sounded at four o'clock on the twenty-third. The thermometer was below zero. There was a foot of snow upon the ground, and it was still coming down furiously when Custer reported to Sheridan that he was ready to move.

"What do you think of this?" asked Sheridan, alluding to the weather.

"It's all right," answered Custer, cheerfully; "we can move. The Indians can't."

There was a hasty breakfast, coffee and hardtack, each trooper standing by the head of his horse, and the column moved off. The undaunted band of the regiment, surely made up of the most heroic and hardy musicians that ever tooted horn or thumped sheepskin, in gallant style played them out and into the terrible blizzard then raging, with the old marching tune "The Girl I left Behind Me," which was more fancy than truth, for there were no "girls" with that expedition, save one hard-featured old campaigner, red-headed at that, who went along as the commanding officer's cook at her own earnest request.

No one can realize the force of a blizzard on the plains who has not, as I have, experienced it. The guides almost immediately declared themselves unable to lead the regiment. Every cavalry officer in the field carries a pocket compass. Custer knew where he wanted to go. With his own compass to show the way he led the regiment forward. The men stumbled on through the awful snow and hurricane until two o'clock, when they were stopped on the bank of Wolf Creek, fifteen miles from the starting point. First caring for the exhausted horses, they made camp, and as the wagons came up fires were soon burning, meals were prepared, and some of the effects of the deadly cold were dissipated.

The next morning, November 24th, they marched down Wolf Creek. The snow had stopped falling, but the temperature stood at seven degrees below zero. The

25th they continued the march. Many another commander would have been stopped by the fearful weather; but Custer was known as a man who would press on as long as the mules could draw the wagons, and when they could not he would abandon the wagons and live off the mules. He kept on. On the twenty-sixth, Thanksgiving Day, arriving at the north bank of the Canadian River, he despatched Major Elliott, the second in command of the regiment, with three troops on a scouting expedition up the river, which he proposed to cross with the balance of his men. There was no Thanksgiving dinner awaiting them, and the remembrance of the holiday spent under happier circumstances but aggravated their present condition.

The river was frozen, but not sufficiently so to bear the regiment. They had to break through the ice and find a ford in the icy water, and it was after eleven o'clock in the morning before the whole command succeeded in passing to the south side. Scarcely had they done so when they noticed a horseman galloping at full speed toward them on the other side. As soon as he came near they recognized him as Scout Corbin, one of Elliott's guides. He brought the startling news that Elliott had come upon the trail of an Indian war party, at least one hundred and fifty strong, and not twenty-four hours old, which led to the south side of the river. The scout was given a fresh horse and ordered to return to Elliott, who was directed to follow the trail cautiously until eight o'clock at night, at which time he was to halt and wait for Custer, who would leave the wagon train and follow him immediately.

Calling the officers to him, Custer briefly gave his orders for the advance. The wagon train was to be left under the care of an officer and eighty men. Each trooper was to take one day's rations of coffee and hard-tack and one hundred rounds of ammunition upon his person, together with a little forage for his horse, and the regiment was to push on at the highest possible speed to join Elliott.

When it came to designate an officer to remain with the train, the detail fell upon Captain Louis McLane Hamilton, whose turn it was to act as officer of the day in camp. This young man bore two historic names. McLane was the second in command of Light Horse Harry Lee's famous cavalry in the Revolution, and he was the grandson of the great Alexander Hamilton. He demurred bitterly to being left in the rear in command of the train under such circumstances. There was no help for it, however, until Custer finally informed him that if he could get any one to take his detail he could go.

It was discovered upon inquiry that one of the officers of the regiment had become almost helpless from snow blindness, the glare of the ice and snow being something terrible, especially upon an open prairie such as they were then traversing. This officer was entirely unfit for active campaigning, but such was his zeal to go forward that he concealed his ailment until Hamilton's scrutiny brought it forth. To him, therefore, was committed the charge of the wagon train, much against his wish, and Hamilton was allowed to go at the head of his troop.

III. The Trail in the Snow

It had grown somewhat warmer during the day. The top crust of the snow became soft, and the horses sank through it to their knees. There was no road or trail, of course, but the command advanced straight across the open prairie toward the point where Corbin had indicated that Elliott had picked up the trail. The several troops were successively placed in the advance for the fatiguing and arduous labor of breaking up the road. There was every desire to spare the horses, but they were nevertheless urged to the last limit to overtake Elliott. Under such circumstances it was problematical whether they would find him alive; for the Indians, who were believed to be in great force, might discover him, ambush him, attack him, and wipe him out as Fetterman had been annihilated, or as Forsyth had been overwhelmed.

During the afternoon Custer and his command struck Elliott's trail, but it was not until nine o'clock at night that they overtook him. They found him encamped on the banks of a litttle stream and thoroughly concealed in the timber. With relief the regiment halted, and taking advantage of the deep ravine through which the creek ran, they managed to build a few fires, which, being well screened, were invisible a short distance away. Over the fires the men made coffee, which, with the hardtack, constituted their only meal since morning — a Thanksgiving dinner indeed.

Elliott had followed the trail, which was still well defined, until eight o'clock, and then had halted in accordance with the orders of Custer, and had waited for his commander. A hasty council was held and some were for taking up the advance at once. But it was pointed out that the moon would rise in one hour and by waiting they would have the benefit of the moonlight in following the Indian trail. Besides, the short rest would do the command good. Saddles were taken off, the horses rubbed down and sparingly fed from the scanty supply of forage. At ten the march was once more resumed in this order:

First of all, riding some distance ahead of the main body, were two Osage Indian scouts. One of these was Little Beaver, who was chief of a small band of Indian auxiliaries which had volunteered for the campaign. Next to them came other Indians, several famous frontiersmen, California Joe and Scout Corbin, and a hideous half Negro, half Indian interpreter whose name was Romero, but whom the soldiers facetiously dubbed Romeo, because he was so ugly; then General Custer and his staff, and then, some distance in rear, the successive troops of the regiment in a column of fours. About three miles from their camping place Little Beaver came back to Custer in considerable agitation and declared that he smelled fire. Nobody else smelled anything, but at his insistence the command was halted, and he and one of his men went forward with Custer and one or two of the scouts until they had gone a mile from the halting place.

Sure enough, after surmounting a little hill, they saw ahead of them and some distance away the embers of a fire. The advance party halted. Little Beaver and the other Indians snaked their course over the ground, taking advantage of every cover to learn what they could. With beating hearts the general and the others watched them. Would they stumble upon the foemen then and there? They waited, concealed beneath the hillock, until Little Beaver returned to tell them that the fire had evidently been kindled by the boys guarding the herds of ponies during the day. At any rate it had almost gone out, no one was there, and the way was safe for the present, although the main camp was probably not far distant.

Orders were sent back to the regiment to advance but to keep its present distance behind Custer and the scouts. The command proceeded with the utmost caution, with an excitement in their veins at the stealthy approach with its possible consequences which made them almost insensible to the frightful cold. About half after twelve o'clock on the morning of the twenty-seventh, Custer saw the leading Indian suddenly sink down behind a hill and wave his hand quickly backward. The whole party dismounted, and the commanding officer with one of his scouts crawled to the hill where the Indian lay. Whispering a word or two, Little Beaver pointed straight in front of him.

Half a mile away a huge black blotch was tremulously moving on the snow in the moonlight. Experienced eyes recognized a herd of ponies. Where the ponies were there were the Indians also. Custer watched the scene for a moment, and upon the still air — the wind had died and the night though bitter cold was intensely quiet — he heard the sound of a bell, evidently tied to the neck of the leader of the herd. Dogs barked, and as they waited they marked the thin, shrill cry of a little child. It was an Indian camp beyond peradventure. Beyond it, among the bare and leafless trees, gleamed in the moonlight the ice-bound shores of a half-frozen river — the Washita.

The general, as tender-hearted a man as ever lived, and as kindly for all his fights, tells us how strangely that infant's cry heard on that bitter winter night moved him, appealed to him. It filled his mind with natural regret that war had to be waged and an attack delivered upon a camp in which there were women and children; but the stern necessities of the case permitted no other course.

Copyright by D. F. Barry

MAJ. JOEL H. ELLIOTT * CAPT. LOUIS McL. HAMILTON *

CAPT. JAMES M. BELL CAPT. J. W. BENTEEN

SOME OFFICERS OF THE SEVENTH CAVALRY IN THE WASHITA EXPEDITION

* Killed in the Battle

The band of Indians under his gaze was that of Black Kettle,* Head Chief of the Cheyennes since the death of Roman Nose, one of the most ferocious and brutal of the Plains Indians. The blood of scores was upon his hands and upon the hands of his followers as well. Torture, infamy, treachery, shame beyond estimation, had stained that band. Even then in the camp there were helpless captives, poor women whose fate cannot be described or dwelt upon.

When Custer had satisfied himself at last that he had found the camp for which he had been searching — which appeared to be a very large one from the number of lodges which they thought they could make out in the distance — leaving the scouts to observe the Indians, he tramped back through the snow to the command, and by messengers summoned the officers about him. Taking off their sabers for the moment, so that their clanking would not betray them, the officers crept to the crest of the hill and made themselves as familiar with the situation as they could by such inspection.

There Custer gave them their final orders. The regiment was divided into four squadrons; Major Elliott, with three troops, G, H, and M, was ordered to circle cautiously to the left and get in the rear of the Indian camp. Captain Thompson, with troops B and F, was directed to make a long detour to the right and join Elliott. Captain Myers, with troops E and I, was commanded to move a shorter distance to the right and take position on the left of Thompson, while Custer himself, with the four remaining troops — Captain Hamilton commanding one squadron, comprising troops A and C, Captain West, another, of troops D and K, with the

* Mo-ke-ta-va-ta. — Letter from Mr. W. H. Holmes, Chief of the Bureau of American Ethnology, Smithsonian Institution.

Osages and scouts and forty sharp-shooters under Adjutant Cook — was to approach the village from the point where they then stood.

Not a sound was to be made, not a shot fired, not a signal given. The attack would be delivered at dawn. When they heard the bugler sounding the charge in the still air of the morning they were to rush in immediately. In order not to impede their movements the men were directed to remove their overcoats and leave them in care of the guard in the rear before the attack was delivered. Then, after hearty handshakes and whispered salutations, the officers assembled their several squadrons and silently started out upon the long detours necessary to enable them to reach their designated positions.

The Indian village was located in the valley of a small river in the Indian Territory, an affluent of the Canadian called the Washita. It was in a deep depression, below the surrounding country, and was well sheltered by trees on the banks of the stream, here easily fordable. By the time all preparations had been made in Custer's own detachment it still wanted some four hours to dawn. The troops with Custer had nothing to do but wait where they were, and a weary, freezing wait it was. So insistent was the general that there should be no noise that he refused to allow the men even to beat their breasts to keep up circulation, or to stamp their feet to ward off the numbing cold. Conversation was forbidden. They were dealing with a warrior who was the most watchful of foemen, with men who could detect an enemy, as the Osage had the fire a mile away, seemingly by instinct. They must take every precaution. The men dismounted and stood uncomplainingly by the side of their horses. Some of them wrapped themselves in their overcoats, and attaching their bridle reins to their wrists, lay down on the ground and actually went to sleep.

About an hour before dawn Custer despatched the last squadron under Captain Myers, who had but a short distance to go, and then, as the first pale grayness of the morning began to steal over the eastern hills and mingle with the moonlight, he gave orders to call the troops to attention. The first sergeants went through the ranks and by a touch of the hand woke the sleeping men. Stiff and numb with the cold, they staggered to their feet, took off their overcoats, left them under the care of a small guard, and mounted their horses. Their sabers had been left behind and they were armed with revolvers and Spencer carbines. The officers quickly formed up their troops and with whispered words placed themselves at the head.

The troops were deployed in line, Hamilton's squadron to the right, West's to the left. Cook's sharpshooters were about forty yards in advance of the left, dismounted, their horses being left with the guard. Some distance in front of all the rest rode Custer. Following him was his bugler. Next to the bugler was the indomitable regimental band. The orders were, in Hamilton's last words, "Now men, keep cool; fire low, and not too rapidly."

The Osages had been somewhat doubtful as to the issue of the attack. They had made medicine, war-painted themselves and arrayed themselves for battle, but with a great deal of trepidation. They expected the white soldiers would be beaten, and they reasoned that in that case their allies would endeavor to purchase their own salvation by surrendering the Osages to the vengeance of their enemies. They determined to take such a position as would enable them to be governed by circumstances in their movements — so they could either fight or fly. They knew the reverence with which the soldiers regarded their flag. Never having been in action with the white man, they concluded that the flag would be kept in a place of safety and if they stood religiously close to the banner they would be in a good position to attack or retreat as circumstances required. Consequently, they rallied on the flag. For once the red man's reasoning led him into trouble, for, as it happened, and as it was to be expected, the flag was in the thick of the fight, and, to give them credit, after they saw their mistake and saw no means of rectifying it, the Osages fought as bravely and as efficiently as the rest.

The command went silently down the hill, making for the center of the valley and the trees where lay the Indian camp. The excitement of the situation was intense. Nobody knew just what he was about to encounter. No one could tell whether the other troops had succeeded in getting within supporting distance or not. But Custer knew his officers, and he, rightly in this instance — alas, that it might not have been so in other cases! — depended upon them. Nearer and nearer the line approached the village. Clearer and clearer came the light from the pale sky. Little, hazy clouds of smoke floated above the tepees under the trees, but aside from that there was yet no evidence of life among them.

However cautiously it was conducted the advance of such a body of men over the snow made a great deal of noise. They had come so near the camp that they could not hope to remain undiscovered another moment. At the instant Custer was about to give the signal a rifle shot was heard on the other side of the camp. At first it was thought to be an accidental discharge from one of the other attacking parties. It was afterward learned that shot was fired by Black Kettle himself, who had heard the noise of the advancing troops, for every squadron had reached its appointed place, and practically at the same time they commenced their advance upon the devoted town. So soon as the crack of the rifle broke upon the still air the bugle sounded the charge.

With the first notes Custer turned to the band. Each trumpeter had his trumpet to his lips, each drummer his drum-sticks in the air.

"Play!" he shouted, and for the first time in action the stirring notes of the tune now peculiar to the Seventh Cavalry as its battle music—"Garry Owen"—broke on the air. Three answering bugle calls rang out from the different squadrons on all sides of the village. The cavalry charged, the dismounted soldiers advanced on the run. They all cheered.

IV. The Attack in the Morning

The village was strung along the banks of the creek and the troopers fell upon it like a storm. The Indians, completely surprised, nevertheless did not lose a moment. They poured out of the lodges, and seeking the shelter of the trees or standing knee-deep in the icy water of the river, with the banks acting as rifle-pits, returned the fire of the white men. A few of them succeeded in breaking away, but most of them had to fight where they were, and right well they fought.

Brave Captain Hamilton, who had sought the detail with such zeal, was shot from his horse and instantly killed. Captain Barnitz received a wound through the breast under his heart. Here and there others fell.

Strict orders had been given to spare the women and children. Most of the squaws and children remained hidden in the tepees. Others took part in the defense. The various troops scattered throughout the village and the fighting was hand-to-hand of a most vigorous character. Captain Benteen, galloping forward, was approached by an Indian boy about fourteen years of age on horseback. The boy was armed with a revolver. As the captain drew near he called out to the lad that his life would be safe if he would throw away his weapon. Fearing he could not understand him he made peace signs to him. For reply the boy leveled his weapon and shot at the captain. The bullet missed him. The Indian fired a second time and the bullet cut through the sleeve of Benteen's coat. The captain was still making signs of amity and friendship when the boy fired a third time and hit his horse. As he raised the pistol to fire a fourth time the officer was forced to shoot him dead.

One squaw seized a little white boy, a captive, and broke for the river. She got into hiding in some underbrush where she might have remained unmolested, but such was her malignity that she busied herself by taking pot-shots at the galloping troops with her revolver. They captured her when her revolver was empty and then discovered that she had been fighting them in spite cf a broken leg.

The Indians rallied in certain places favorable for defense. In their desperation seventeen braves threw themselves into a little depression in the ground and refused to surrender, fighting until all were killed. In a ravine running from the river thirty-eight made a heroic defense until they were all shot. In all, one hundred and three were killed, including Black Kettle, the chief.

The furious fighting had lasted one hour. The village was now in possession of the troops. A number of officers and men had been wounded and a temporary hospital was established in the middle of the village. Details were sent through the lodges to rout out the squaws and children, and a roll-call was ordered.

Custer was dismayed to find that Major Elliott and fourteen men, including Sergeant-Major Kennedy and three corporals, were missing. Where they had gone to no one at first could imagine. Finally a trooper stated that a number of Indians had escaped in the gap bettween Elliott and Thompson, and that he had seen Elliott with a few troopers break away in pursuit of them. An order was given for a troop to search for them, but before it could get away Indians were perceived in a heavy force on the bluffs directly in front of the command. Custer had succeeded in killing practically the whole of Black Kettle's band, and as the Indians who had escaped had been forced to run for their lives, naked as they came from the lodges, he could not understand the appearance, just out of range of his men, of this portentous and constantly increasing force arrayed in full war panoply.

Inquiry among the captives disclosed the fact that the valley had been chosen as the winter headquarters for the principal bands of the Kiowas, Arapahoes, Cheyennes, "Dog Soldiers,"* Comanches, and even a wandering band of Apaches. There were at least two thousand warriors in this assemblage. At that moment

* Dog Soldiers were bands of especially ruthless Indians who could not brook even tribal restraint. They included members of different tribes and were unusually formidable. Possibly they got their name from a perversion of Cheyennes, *i.e.*, Chiens-dogs. Another account describes them as a sort of mercenary police at the service of a chief of a tribe, with which he enforced his commands upon the recalcitrant and generally kept order. In any case they were men of exceptional courage and bravery.

the men who had been guarding the overcoats and the lead horses came running in saying that they had been driven off by a heavy force of Indians. The situation was indeed critical.

Something had to be done at once. Custer dismounted his men, threw them out in a half circle about the camp, and prepared for battle. The Indians did not delay in delivering it. Led by Little Raven, an Arapahoe, and Satanta,* a famous Kiowa, and Black Kettle's successor, Little Rock, they at once attacked. A fierce battle was on and Custer's ammunition was running low. The troops were now fighting for their lives. They had not expected anything of this kind. Fortunately, at this critical juncture a four-mule wagon came dashing through the Indian line. The Indians, being occupied in fighting, did not observe it until it was right upon them. Driving the wagon was Major Bell, the quartermaster, from the train. With him was a small escort. He had loaded the wagon with ammunition and galloped toward the sound of the fighting. With the fresh supplies, therefore, the troops at last made a bold charge which drove the Indians headlong down the valley, during which Little Rock, striving to rally his braves, was killed.

Custer now set fire to the lodges, totally destroying them and their contents. What to do with the ponies of the herd which had been captured in spite of the efforts of the squaws to run off with them, was a problem. It was impossible, under the circumstances, to drive them back to the camp. To turn them loose would have allowed them to fall into the hands of other Indians for use in future warfare. They had to be shot. It

*A corruption of Set-t'á-iñt-e, "White Bear."— Letter from Mr. W. H. Holmes, Chief of the Bureau of American Ethnology, Smithsonian Institution.

was a most unpleasant and repulsive duty for the soldiers, but there was no alternative. The whole herd was slaughtered. It took an hour and a half to kill them, and those engaged in the work said they had never done anything so harrowing and distressing.

By this time it was late in the afternoon. The Indians from the other villages, finding they were pursued but a short distance, had reassembled and once more prepared for attack. It was necessary for Custer to retreat at once. He put every available man on horseback, threw out skirmishing parties, the colors were brought up, the indefatigable band started playing, and the party advanced gaily up the valley toward the Indians. As he hoped and planned, they immediately reasoned that he would not advance with such confidence against such an overwhelming force, unless he was supported by heavy reinforcements to his command. After a short resistance they broke and fled.

It was night by this time, and Custer lost no time in getting out of the valley. The weather was still frightfully cold, and his men were without their overcoats, for they had, of course, not recovered them, and were almost perishing. They got back in safety, however, to Camp Supply, having accomplished the object of their expedition in dealing a decisive blow to the Indians. More than that, they had shown the Indians, who trusted for immunity to the season, that winter and summer were alike to the American soldier.

The Indian loss was one hundred and three killed in the village, including Black Kettle; an unknown number, believed to be large, killed and wounded during the all-day fighting, including Little Rock; the capture of fifty-three squaws and children; eight hundred and seventy-five ponies, eleven hundred and twenty-three buffalo robes and skins; the destruction of over five hundred pounds of powder and one thousand pounds of lead; four thousand arrows, seven hundred pounds of tobacco, besides rifles, pistols, saddle-bows, lariats, immense quantities of dried beef, and other winter provisions; in short, the complete destruction of the village and the annihilation of the band.

The losses of the regiment in the engagements were one officer and fourteen men missing (Elliott and his party), one officer and five men killed, three officers and eleven men wounded. General Sheridan called the affair the most complete and satisfactory battle ever waged against the Indians to that time.

Custer had marched through that blizzard and over the snow-clad plains to victory. His stealthy approach, the skill with which he had surrounded the village, the strength with which the attack had been delivered, and the battle which he had fought with the unexpected Indian force, the ruse by which he had extricated himself, and, last but not least, Bell's gallant dash with the ammunition wagon, were all given the highest praise. And well they merited it.

One or two incidents of the battle are worthy of especial mention. When the troops obtained possession of the village, they found the dead body of a white woman. The fact that she still had some vestige of civilized clothing upon her, quite new, proved that she had been but recently captured. She had been shot dead by the Indians at the moment of attack to prevent her rescue; and there was also the body of a little white child, who had been killed by those who had him in charge, lest he should be returned to his family again.

The squaws, of course, were in great terror. They thought they would be instantly put to death when they were routed from their tepees. Black Kettle's sister, Mah-wis-sa, who seemed to be the leading woman of the village, made a long oration to Custer, telling him that she was a good Indian, and that she had tried to restrain Black Kettle in his nefarious career — which was all a lie, of course. She wound up by bringing the comeliest of the young Indian maidens to Custer, and, after solemnly placing the hand of the girl in that of the General, mumbled some kind of a gibberish over the two. The General observed Romeo standing near with a broad grin upon his face, and asked him what Mah-wis-sa was doing. He was told that she was marrying him to the beauty of the tribe to propitiate him. That marriage did not stand.

V. The Fate of Elliott and his Men

The fate of Elliott's detachment remained a mystery. His comrades hoped that he had escaped, but as the days passed and he did not return to the regiment, and as nothing was heard from him, they abandoned hope in despair. This was not, by any means, the end of the winter campaigning; and some time after, Custer and his men, this time heavily reinforced, again marched up the valley of the Washita. A short distance from the place where Black Kettle's band had been annihilated they found the remains of Elliott and his men. The evidence of the field and what was afterward learned from Indian captives told the sad story.

Pursuing the fleeing Indians, Elliott and his party suddenly ran into the midst of a horde of braves coming down the valley to help Black Kettle and the men who had been engaged with Custer. To fly was impossible. They dismounted from their horses, formed themselves in a semicircle a few feet in diameter, stood back to back, as it were, and fought until they died. There were evidences of a terrible conflict all around them. Right dearly had they sold their lives.

The last survivor of that gallant little band had been Sergeant-Major Kennedy, the finest soldier in the regiment. He was not wounded, it appeared, but had expended all his ammunition for both rifle and revolver. Being an officer, he wore a sword. Seeing him, as they supposed, helpless, the Indians resolved to take him alive for the purpose of torturing him. There was not a soldier who knew of the habits of the Indians who would not chose death to captivity any time. The brave Kennedy stood alone in the midst of the bodies of his comrades, fronting death, sword in hand. I like to think of the courage of that heroic man in the midst of that savage, ravening assemblage.

With wily treachery the Indians made peace signs, and walked toward him with hands outstretched, saying: "How, How!" Kennedy, who knew the true value of such proceedings, waited until the chief of the band approached him nearly, then thrust his sword up to the hilt into the Indian's breast. When they found Kennedy's body he had been pierced by no less than twenty bullets. The other troopers had received one or two bullet wounds each. They were all stripped, scalped, and mutilated.*

There was a great outcry when this battle became known, and Custer was accused of slaughtering helpless, inoffensive, gentle Indians! Unmerited obloquy was heaped upon him, but those who lived near enough to feel the effects of the Red Scourge realized that he had done for the settlers the best thing that could be done.

* "Our Wild Indians," Colonel Richard I. Dodge, U. S. A.

People who knew, and his superior officers, not only sustained but commended him.

Custer again, in command of a much larger force, surprised a more populous village later in this same winter. It was completely in his power. He could have wiped it from the face of the earth, although it contained a force of Indians nearly equal to his own; but he stayed his hand, and said he would spare the savages if they would deliver to him two wretched women, one a young bride, the other a young girl, whom they held in captivity and for whose deliverance the campaign had been undertaken. By masterly skill Custer captured Satanta the Infamous, and held him until the captives were given up. With the expedition was the brother of one of the captives. Custer tells, in his simple, terse manner, with what feelings that whole army watched the poor women brought into camp, and how the boy, the last of his family, stood trembling by the general's side until he recognized, in one of the wrecks of humanity which the Indians handed over, the sister whom he was seeking.* The red-headed cook, referred to above, was with the army again, and proved herself, in her rude way, an angel of mercy and tenderness to these, her wretched sisters.

* " . . . and at the last the brief reference to that episode when he (Custer) let glory of battle go, to save two white women!

"Has any one told you that the long line of soldiers and officers drawn up to witness the return of the two captives wept like women, and were not ashamed when the poor creatures came into the lines? Will you not write that story up some day, Dr. Brady? I will give you some addresses of officers who were eye-witnesses. They cannot seem to put such a picture before the public, but they do talk well."—Private letter to me from the wife of an officer present on the occasion noted.

CHAPTER ELEVEN

Carr and Tall Bull at Summit Springs

I. A Brilliant Little Fight

GENERAL Eugene A. Carr, in command of the Fifth Cavalry, did some brilliant skirmishing and fighting in 1868-9 in western Kansas and Colorado. His most notable exploit was the surprise of Tall Bull's camp. Next to Black Kettle, Tall Bull was probably the most vicious and diabolical of the Indian raiders in these two states.

Carr, with five troops of the Fifth Cavalry and with W. F. Cody (Buffalo Bill) as chief guide, learning where Tall Bull's camp was, marched one hundred and fifty miles in four days in pursuit of him. Halting when he believed he had reached the vicinity of the camp, he sent Buffalo Bill with some of his Pawnee Indian auxiliaries to find out exactly where the Indians were located.

Cody, having discovered the location of the village, returned to General Carr and advised him to take a wider detour, keeping his forces concealed among the hills, so that he could approach the Indians from the north, a direction from which they would not be expecting attack, and whence they might be the more easily surprised. The advice was followed, the command made its encircling march without detection, and formed up in line of troops, each troop two abreast, in the ravines about twelve hundred yards from the village.

They were between the Indians and the Platte River. The camp was located at Summit Springs, Colorado. Every preparation having been made, Carr ordered the bugler to sound the charge. The man was so excited that he was unable to produce a note. Twice Carr gave the command. Finally, Quartermaster Hayes snatched the bugle from the agitated musician and sounded the charge himself, and the whole regiment rushed out into the open.

The Indians made for their ponies and advanced to meet the charge. The rush of the soldiers was too threatening, however. After a hasty fire they broke and fled on their horses, the whole party, soldiers and Indian scouts, following after at full speed through the village. The attack was a complete success. Fifty-two Indians were killed, two hundred and seventy-four horses and one hundred and forty-five mules were captured. The soldiers had one man wounded, with no other casualties.

In the camp were found the bodies of two unfortunate white women, who had been captured. Swift as had been the dash of the soldiers, the Indians had taken time to brain one of the women with a war-club, while the second was shot in the breast and left for dead. She was given every possible attention by the soldiers, who took her back to Fort Sedgwick, and her life was eventually spared. Her sufferings and treatment had been beyond description. Fifteen hundred dollars in money — in gold, silver, and greenbacks — strange to say, had been found in the camp. This sum the soldiers, by permission of the general, donated to the poor woman, as an expression of their sympathy for her.

According to some accounts, Tall Bull, who was chief

of the camp, and one of the head chiefs of the Sioux, was killed in this attack. Buffalo Bill tells another story.* The day after the fight the various companies of the Fifth Cavalry — which had remained in the camp all the ensuing day and night, at the insistence of the plucky commander, in spite of the pleas of some of the officers, who, fearing an attack in force, suggested retiring immediately — separated in order the more effectively to pursue the flying Indians. Several days after the surprise the detachment for which Cody was guide was attacked by several hundred Indians. The soldiers fought them off, killing a number. The chief of this party was believed by Cody to be Tall Bull.

Buffalo Bill crept through a ravine for several hundred feet, unobserved by the Indians, until he reached an opening whence he had the savages in range. Watching his opportunity as the Indians were careering wildly over the prairie, he drew a bead on the chief and shot him dead. Whether that was Tall Bull or not, one fact is clear — that he was killed either then or before, for he was certainly dead thereafter.

When the troops were following the Indian trails on the march to Summit Springs, at every place where the Indians had camped they found marks of white women's shoes. It was this knowledge that gave additional determination and fire to their magnificent attack.

General Carr deservedly gained great reputation for his dash and daring.

Here I include a letter describing this battle from the standpoint of a soldier, which is a most interesting contribution to the story of the affair:

* I have written several times to General Carr, asking information as to this and other points, but have not received any.—C. T. B.

II. Account of the Battle of Summit Springs

Written by J. E. Welch to his comrade, Colonel Henry O. Clark, of Vermont.*

The next spring, 1869, I heard that General Eugene A. Carr, commanding a detachment of the Fifth U. S. Cavalry, was organizing an expedition to go after a large band of Indians (Sioux and Cheyennes) who had been raiding and murdering through Colorado, New Mexico, and Kansas. Some other fellows and myself went to Fort McPherson and offered our services as volunteers to serve without pay. The general could not accept our services, but he said we could go along and act with the scouts — so along we went.

The expedition consisted of about four hundred cavalry, one hundred and fifty Pawnee scouts, under Colonel Frank North, and about twenty civilians. Buffalo Bill was the guide. He struck out for the Republican River, and the first night after we got there the hostiles tried to stampede our horses; they came near accomplishing their object, too, but they only succeeded in wounding a teamster and killing a mule. Next day we found the trail of their main body and followed it, but soon found that we could not gain an inch on them; we kept on, however, until we came to a place where the trail divided. The trail to the right was very plain, while the one to the left was scattered and so dim it was evident to the most inexperienced man in the command that the trail they intended us to follow was the one to the right. So General Carr detached two troops of

* This letter, which is dated Edith, Coke County, Texas, June 16, 1891, was furnished me for publication by Dr. T. E. Oertel. I am informed that the writer has since died.—C. T. B.

cavalry and some Indians, under Major Royall, caused them to make as big a show as possible and take the decoy trail, while the main body was kept back in a low place for a day in order that any hostile scouts who might be watching us would think the whole command had gone on the decoy trail. Next day we started on the dim trail, and before night we became satisfied that we were on the trail of the main body of the hostiles. Major Royall followed the decoy trail until it scattered, then turning the head of his column to the left he intercepted and rejoined the command. We now found that we were gaining on the game we were after. They evidently thought they had fooled us, and were taking their time.

On the tenth of July we marched sixty-five miles, passing three of their camps. On the eleventh we were on the march before daylight. The trail was hot, the Indians making for the Platte. Every one knew that if they succeeded in crossing the river the game was up. By noon we had marched thirty-five miles, at which time Buffalo Bill, who had been far in advance of the command all day, was seen approaching as fast as his tired horse could come. As soon as he reached the column he called for a fresh horse, and while transferring his saddle told General Carr that he had encountered two bucks who were hunting and that the Indian camp was about twelve miles ahead.

The general, knowing the bucks who had been run off by Cody would make every effort to reach their camp ahead of us in order to give the alarm, gave the command "Trot." Both horses and men seemed to brighten up, and we put real estate behind us at a rapid rate. When within a mile of the hostile camp a halt was called to let the Pawnees unsaddle, as they flatly refused to go

GEN. GEORGE CROOK
GEN. ELWELL S. OTIS

GEN. EUGENE A. CARR
GEN. HENRY B. CARRINGTON

GROUP OF DISTINGUISHED GENERAL OFFICERS

into action with saddles on their horses. They began daubing their faces with paint and throwing off their clothing. They were made to retain enough of the latter to enable us to distinguish them from the hostiles. After this short delay we moved forward at a sharp trot, and in a few moments we were looking down at "Tall Bull's" camp in a small valley below us. In a moment the camp was alive with Indians running in every direction.

General Carr, taking in the situation at a glance, gave utterance to a few words of command, winding up his remarks with the order, given loud and clear and sharp:

"Charge!"

Every horse leaped forward at the word, and in a twinkling we were among them and the fight was on. It did not last long. There was rapid firing for about five minutes, when all was over except an occasional shot as some fellow would find an Indian who had failed to secure a horse and escape.

The result of the fight was about as following: no white men killed, four or five horses killed, about one hundred and eighty-eight dead Indians, forty of whom were squaws and children;* one hundred and five lodges captured, many rifles, five tons of dried buffalo meat baled for winter use, a very ample supply of ammunition, consisting of powder, lead, etc., and a greater number and variety of brass kettles than I ever saw before.

Of their live stock we captured five hundred and sixty head of ponies and mules.

To pursue those who had fled was out of the question, our horses being too badly done up. As we charged the camp, we saw a white woman run from among the

* These figures, which are evidently from memory, are certainly in error.—C.T.B.

Indians, one of whom fired at her as she ran. We shouted to her to lie down, which she did, our horses leaping over her without a hoof touching her. She was wounded in her side, but not fatally. Almost at the same moment we saw an Indian seize another white woman by the hair and brain her with a tomahawk. Some of us rode straight for that Indian, and there was not a bone left in his dead carcass that was not broken by a bullet. I dismounted in the midst of the hubbub to see if I could help the woman, but the poor creature was dead. (She had the appearance of being far gone in pregnancy.) I mounted my horse again with a very good stomach for a fight.

After firing a few shots, I happened to see a Red mounted on a large paint pony making off by himself, and driving four fine mules ahead of him. I gave chase and gained on him rapidly, which he soon perceived, dropping his mules and doing the best he could to get away. But it was no use. "Sam," my horse, was Kentucky bred, and walked right up on him. When I was within seventy-five or one hundred yards of him he wheeled his horse and fired, the bullet passing through the calf of my leg and into my horse. The Indian threw his gun away and rode at me like a man, discharging arrows as he came. The third arrow split my left ear right up to my head. It was then my turn, and I shot him through the head. This Indian's name was "Pretty Bear." He was chief of a band of Cheyennes. The Pawnees knew him and were anxious to secure his scalp, which I was glad to give them as I soon became disgusted with the ghastly trophy. "Pretty Bear" had on his person the badge of a Royal Arch Mason, with West Springfield, Ill., engraved on it. I sent the badge to the postmaster at Springfield with a statement as to how it came into my possession. "Pretty Bear" had five or six scalps on the trail of his shield, one of which was that of a woman. The hair was brown, very long, and silken.

"Tall Bull," the Sioux chief, was killed by Lieutenant Mason, who rode up to him and shot him through the heart with a derringer. After I had taken the scalp of "Pretty Bear" I found that Sam was shot through the bowels. I unsaddled him and turned him loose to die, but he followed me like a dog and would put his head against me and push, groaning like a person. I was forced to shoot him to end his misery. I had to try two or three times before I could do it. At first to save my life I could not do it. He kept looking at me with his great brown eyes. When I did fire he never knew what hurt him. He was a splendid horse, and could do his mile in 1.57.

My wounds being slight, I rustled around and soon managed to catch a small mule, which I mounted bareback, intending to scout around a little. I did not carry out my intention, however. The brevet horse ran into the middle of the Indian camp, threw me into a big black mud-hole, my boot was full of blood, my ear had bled all over one side of me, so that when I crawled out of that mud-hole I was just too sweet for anything. By this time the fight was over. A friend of mine, Bill Steele, went with me to the spring that ran into the mud-hole, where he washed me as well as he could, bandaged my leg, sewed my ear together with an awl and some linen thread. He made a good job of it, and I was all right except that my leg was a little sore and stiff.

After the fight we found we had one hundred and seventeen prisoners, four squaws, and fifteen children. They were turned over to the Pawnees. The Pawnees did not fight well. They skulked and killed the women and children. I have never seen Indians face the music like white men. We camped where we were that night. Men were coming into camp all night. In fact, they did not reach the scene of action until about ten o'clock next day. They were fellows who had been left along the trail by reason of their horses giving out.

Our first duty next day was to bury the poor woman they had so foully murdered the day before. Not having a coffin, we wrapped her in a buffalo robe. General Carr read the funeral service and the cavalry sounded the funeral dirge, and as the soft, mournful notes died away many a cheek was wet that had long been a stranger to tears. The other woman was found to be all right with the exception of a wound in the side. She was a German, unable to speak English. Both of the women had been beaten and outraged in every conceivable manner. Their condition was pitiful beyond any power of mine to portray.

The Indian camp and everything pertaining thereto was destroyed, after which we took up our line of march for Fort Sedgwick, where we arrived in due time without any mishap.

I think it just as impossible to make a civilized man of the Indian as it would be to make a shepherd dog of a wolf, or a manly man of a dude. They do not in my opinion possess a single trait that elevates a man above a brute. They are treacherous, cowardly, and ungrateful, Cooper to the contrary notwithstanding. I knew a Greek in Arizona who came to the country with camels for the government. After the camels died he married an Apache squaw, learned the language, and was employed by the United States government as an interpreter. This man told me that in the Apache dialect there was no word, or combination of words, whereby they could convey the idea that we do by using the word *Gratitude*. What do you think of that?

Well, old man, I have been writing half the night, and have only got as far as the 11th of July '69. I am discouraged, and right here I quit you like a steer in the road. How long am I to wait for that picture? I am curious to see how much of a change old dad Time has wrought in you. He has played h—l with me.

As ever,

J. E. WELCH.

P. S. The photo has come. I could have known you anywhere. You have changed a little — for the better, I think.

J. E. W.

Part II

The War With the Sioux

CHAPTER ONE

With Crook's Advance

I. The Cause of the Fighting

LATE in 1876 the government determined that thereafter all Indians in the Northwest must live on the reservations. For a long time the Interior Department, to which the management of Indian affairs was committed, had been trying in vain by peaceable means to induce them to do this. The Indians were at last definitely informed that if they did not come into the reservation by the first of January, 1876, and stay there, the task of compelling them to do so would be turned over to the War Department. They did not come in; on the contrary, many of those on the reservations left them for the field; and thus the war began.

The principal adviser and most influential head man among the Sioux Nation and its allied tribe, the Cheyennes, was Sitting Bull,* an Unkpapa chief and a great medicine man. He does not seem to have been much of a fighter. The Indians said he had a big head but a little heart, and they esteemed him something of a coward; in spite of this, his influence over the chiefs

* Tatá ^{n}ka I yotá ^{n}ka, according to a letter from Mr. W. H. Holmes, Chief of the Bureau of American Ethnology, Smithsonian Institution.

and the Indians was paramount, and remained so until his death.

Perhaps he lacked the physical courage which is necessary in fighting, but he must have had abundant moral courage, for he was the most implacable enemy and the most dangerous — because of his ability, which was so great as to overcome the Indians' contempt for his lack of personal courage — that the United States had ever had among the Indians. He was a strategist, a tactician — everything but a fighter. However, his lack of fighting qualities was not serious, for he gathered around him a dauntless array of war chiefs, the first among them being Crazy Horse, an Oglala, a skilful and indomitable, as well as a brave and ferocious, leader.

The Sioux country was encircled by forts and agencies. The Missouri River inclosed it on the east and north. On the south were the military posts along the line of the Union Pacific Railroad. To the west were the mountains. Sitting Bull and his followers took position in the valleys of the Big Horn near the sources of the Powder River, right at the center of the encircling forts and agencies. It was a situation whence they could move directly upon the enemy in any direction as necessity required.

For years unscrupulous and mercenary traders had supplied the savages with high-grade firearms in spite of government protests. The Indians were better armed than the soldiers, and possessed ammunition in plenty. Their numbers in the field have been estimated at from twelve hundred to six thousand warriors, with their wives and children. Those who have studied the war from the Indian point of view have put the number at the lower figure; nearly every one else at from three thousand up. Whatever the facts, there were enough of them to give the United States Army the busiest time that it had enjoyed since the Civil War.

Three expeditions were planned for the winter, which were to be launched upon the Indians simultaneously. One, under General Gibbon, was to come eastward from western Montana; another, under General Crook, was to advance northward from southern Nebraska; and the third, under Custer, was to strike westward from Fort Lincoln. It was believed that any one of the three, each of which comprised more than a thousand men, would be strong enough to defeat the Indians, the only problem being to catch them or corner them.

The well-known disinclination of Indians to fight pitched battles is a factor which enters largely into every campaign. Somehow or other, the Indians in this campaign did not seem to be so disinclined that way. One cannot but admire the skill with which they manœuvered and the courage with which they fought. Putting aside all questions of their cruelty and brutality — and what else could be expected from them ? — they were patriots fighting for the possession of their native land. Bravely they fought, and well. They were fully apprised of the movements of the troops, and resolved to attack them in severalty and beat them in detail. We shall see how completely they did so, and with what brilliant success they battled, until they were run down, worn out, scattered, killed, or captured.

II. Reynolds' Abortive Attempt

The weather was something frightful. Indeed, all through the ensuing spring it was unprecedentedly inclement. Neither Custer's expedition nor Gibbon's got away in winter. Crook did advance, and first came in touch with the enemy with results not altogether satisfactory. General Joseph J. Reynolds, with ten troops from the Second and Third Cavalry, surprised and took possession of Crazy Horse's village, on the Powder River, on the morning of March 17, 1876. The troops had partially destroyed the village while under a severe fire from the Indians who had rallied on bluffs and hills round about it, when Reynolds abandoned the position and retreated. He was, of course, pursued by the Indians, grown bolder than before, if possible, as they saw the reluctant soldiers giving up their hard-won prize.

So precipitate was Reynolds' withdrawal, in fact, that the bodies of several troopers who had been shot in the action were abandoned to the malignity of the savages, and there was a persistent whisper, which will not down, to the effect that one wounded man was also left behind.

As to this, an army officer of high rank personally stated to me that Reynolds was in such a state of excitement, as the afternoon wore away and Crook did not join him in the village, that he finally peremptorily ordered the troops to mount and go away, in spite of the fact that the work of destruction was not complete. This was bad enough, but my informant solemnly asserted that Reynolds, in spite of plea and even remonstrance, compelled him to leave behind a wounded trooper, who must necessarily have been tortured by the Indians so soon as they re-occupied the village. Captain Bourke has gone on record in his "On the Border with Crook," expressing his belief in the truth of this charge, which forever stains the name of the commander of the expedition. The whole affair was a disgrace to the army, and many of the officers of the command, capable and brave men, felt it keenly. They chafed for a chance to show their qualities, which they had later on.

The cold was intense, the temperature dropping to thirty degrees below zero. The soldiers suffered greatly in the retreat. The Indians, who seemed impervious to cold, pursued them and succeeded in recapturing their pony herd of some seven hundred head, which Reynolds was endeavoring to bring away with him. Crook, bringing up the infantry and wagons, was furious when he met the retreating cavalry and heard its story.

There were a number of courts-martial subsequently, but little came of them, and the matter was finally allowed to drop upon the retirement or resignation of some of the officers chiefly concerned. It was a disgraceful affair, and all the honors rested with Crazy Horse. The Indians were greatly encouraged. The loss of the troops was four men killed and six wounded, and sixty-six men badly frozen or otherwise incapacitated by the cold.

III. The March to the Tongue River

After the ignominious outcome of Reynolds' attack upon the village of Crazy Horse, the various expeditions noted spent the greater part of the spring in preparing for the grand advance of the converging columns which were to inclose the recalcitrant Indians in a cordon of soldiers, force them back on the reservations, and thus, it was sincerely hoped, end the war. It will be necessary to follow the movements of the several columns separately. As that of Crook first came in contact with the Indians, its history will be first discussed.

The reorganized command for the campaign, which assembled at Fort Fetterman, Wyoming, included fifteen troops of cavalry — about nine hundred men — ten of the Third, under command of Colonel Evans, and five of the Second, under Major Noyes, the whole being under the command of Colonel William B. Royall, of the Third.* There were also three companies of the Ninth Infantry and two of the Fourth, a total of three hundred men, under the command of Colonel Chambers.

There was an abundance of transportation, a long wagon train, and an invaluable pack train. The troops were generously provided with everything necessary for the hard work before them. It was the largest, and it was believed to be the most efficient, force which had ever been sent against the Indians in the West.

Crook, an officer of large experience, especially in Indian fighting, assumed personal command of the expedition on the 28th of May, 1876. On the 29th the march began. The objects of the campaign were the

* General King, in his fascinating book, "Campaigning with Crook," has preserved a characteristic anecdote of Royall, which I venture to quote as illustrating the way they have in the army, and as throwing some light on the temperament of the peppery old fighter:

"A story is going the rounds about Royall that does us all good, even in that dismal weather. A day or two before, so it was told, Royall ordered one of his battalion commanders to 'put that battalion in camp on the other side of the river, facing east.' A prominent and well-known characteristic of the subordinate officer referred to was a tendency to split hairs, discuss orders, and, in fine, to make trouble where there was a ghost of a chance of so doing unpunished. Presently the colonel saw that his instructions were not being carried out, and, not being in a mood for indirect action, he put spurs to his horse, dashed through the stream, and reined up alongside the victim with:

"'Didn't I order you, sir, to put your battalion in camp along the river, facing east?'

"'Yes, sir; but this ain't a river. It's only a creek.'

"'Creek be d—d, sir! It's a river — a river from this time forth, *by order*, sir. Now do as I tell you!'

"There was no further delay."

villages of Crazy Horse and Sitting Bull, which were believed to be somewhere on the Rosebud River. The topography of that country is well known now, but then it was more or less of a terra incognita — rather more than less, by the way. Certainly, this was true after the Tongue River was reached. The advance was made at first up the Bozeman trail, past Fort Reno, and over the battle-fields around the ruins of abandoned and destroyed Fort Phil Kearney, which were objects of much interest to the soldiers.

On the 9th of June the army encamped on the south side of the Tongue River, near the point where that stream intersects the Montana boundary line. Crazy Horse * had been fully advised by disaffected Indians at the agencies and military posts, as well as by his own daring scouts, of all these preparations that were being made to overwhelm him. He had sent to Crook a specific warning not to cross the Tongue River, and declared his intent to attack him immediately he reached that stream. To prove that his threat was no idle boast, he mustered his warriors, and at half after six o'clock on the evening of the 9th, from the high bluffs on the other side, opened fire upon the camp.

Through a fortunate mistake the Indians directed their fire to the tents of the camp, imagining that they would be full of men. They happened to be empty. The Sioux soon got the range, and the camp was swept with bullets. They ripped open mess chests,

* "Crazy Horse was the personification of savage ferocity; though comparatively a young man, he was of a most restless and adventurous disposition, and had arrived at great renown among the warriors, even before he was twenty-six years of age. In fact, he had become the war chief of the southern Sioux and the recognized leader of the hostile Oglalas."—"Personal Recollections of General Nelson A. Miles, U. S. A."

shattered the sides of the wagons, destroyed the baggage, killed a few horses, but did little damage to the men.*

The Third Cavalry was divided into three battalions, one of four, and two of three troops each. Captain Mills commanded the first battalion, Captain Henry the second, Captain Van Vliet the third. Crook acted promptly. He sent forward three companies of his infantry, deploying them as skirmishers, to line the river bank and open fire on the Indians in plain view on the bluff on the other side. At the same time he ordered Captain Mills to take his battalion across the river and charge the enemy. The Sioux were already unsettled by the accurate fire of the infantry with their long-range rifles, and as Mills' battalion deployed, dashed through the water and at the steep bluffs on the other side, they broke and fled, having suffered little or no loss, and not having inflicted much more.

IV. The Flying Column

The skirmish was simply a grim earnest of the determined purpose of the Indian chief. No pursuit was attempted at that time. Negotiations had been entered into between the Crows, who were the hereditary enemies of the Sioux, and the Shoshones, with a view to securing a body of Indian auxiliaries to the troops, whose services would be invaluable for scouting. Persuaded

* One bullet smashed the pipe of a small camp stove in Captain Mills' tent. When the Eastern papers learned the interesting fact that Mills' stovepipe had been smashed, that gallant officer was severely censured, and much ridicule was heaped upon him, under the impression that he wore a "stovepipe" hat in action. By the way, when Captain Broke, of the British frigate *Shannon*, boarded the American frigate *Chesapeake*, Captain Lawrence, in the War of 1812, he wore just such a hat!

thereto by Frank Gruard,* a celebrated scout, something less than two hundred Crows, with eighty odd Shoshones, joined the army on the 15th of June.

To pursue Indians while incumbered with infantry and a wagon train was well nigh a hopeless task. Crook determined to park the wagons and baggage, leave them under the command of Major Furey, the quartermaster, strip his command to the lightest marching order, and make a dash for the Rosebud River and the Indian country. One hundred infantrymen, protesting most vehemently against their orders, to their credit, be it said, were detailed to remain with the train. Two hundred others, who professed to have some skill in riding, were mounted on the mules of the wagon train to accompany the cavalry.

The morning of the 15th was spent in accustoming the infantrymen to the mules and the mules to the infantrymen. The cavalrymen and the Indian allies enjoyed the circus which ensued when the mules were bridled and saddled for the first time, and mounted by men who had never before straddled anything more formidable than a fence rail. It took the whole morning before the infantrymen and the mules learned to get along with each other, even in a half-hearted way.

* "Frank Gruard, a native of the Sandwich Islands, was for some years a mail rider in northern Montana, and was there captured by the forces of 'Crazy Horse'; his dark skin and general appearance gave his captors the impression that Frank was a native Indian, whom they had recaptured from the whites; consequently, they did not kill him, but kept him a prisoner until he could recover what they believed to be his native tongue — the Sioux. Frank remained several years in the household of the great chief 'Crazy Horse,' whom he knew very well, as well as his medicine man, the since renowned 'Sitting Bull.' Gruard was one of the most remarkable woodsmen I have ever met; no Indian could surpass him in his intimate acquaintance with all that pertained to the topography, animal life, and other particulars of the great region between the head of the Piney, the first affluent of the Powder on the west, up to and beyond the Yellowstone on the north; no question could be asked him that he could not answer at once and correctly. His bravery and fidelity were never questioned; he never flinched under fire, and never growled at privation."—"On the Border with Crook," Captain John G. Bourke.

At five A.M., on the 16th, the force, numbering a little less than eleven hundred men, with two hundred and fifty Indian auxiliaries, crossed the Tongue River and marched to the Rosebud.* They bivouacked that night on the banks of the Rosebud, on a level depression surrounded by low bluffs on all sides, forming a sort of natural amphitheater, on the top of which the pickets were stationed. Each man carried four days' rations of hardtack, coffee, and bacon in his saddle-bags and one hundred rounds of ammunition upon his person. The pack train was limited to two mules carrying the medical supplies. There being little to do at the wagon camp on Goose Creek, a number of mule packers, led by a veteran, Tom Moore, accompanied the expedition to help the foot soldiers to manage their mules, and incidentally to take part in the fighting. There were no tents, of course, and but one blanket (a single blanket at that) for each man. This blanket barely kept off the heavy dew, and the night was a thoroughly uncomfortable one.

At three A.M. on the 17th of June reveillé was sounded. After breakfast and the care of the horses and mules, six o'clock found the troops on the march down the Rosebud. At eight o'clock they halted and unsaddled their animals to give them a nibble of grass and a little rest, preparatory to a farther advance later on, while the Crows and Shoshones were sent on ahead to scout. The place in which they had stopped was an amphitheater, like their camp ground of the night before, a rolling bit of boggy prairie, inclosed on all sides by bluffs, every point being within rifle shot of the center. Through this amphitheater ran the Rosebud River, here a mere creek, its general direction being

* So called from the quantity of wild roses which grew along its banks in season.

from west to east. Toward the east side of the amphitheater the creek was diverted to the left, the northeastward, and plunged into a gloomy and forbidding cañon, called the Dead Cañon of the Rosebud. The course of the river was marked by a rank undergrowth of grass, small trees, etc.

Mills' battalion of the Third Cavalry halted on the south bank of the creek. In rear of Mills was Noyes' battalion of the Second Cavalry. Across the creek were Henry's and Van Vliet's battalions of the Third Cavalry, the mounted infantrymen, and the small pack train with the packers. Crook desired to keep his movements secret, but it had been impossible to restrain the impetuosity of the Indian auxiliaries the day before. They had come across a herd of buffalo and had made great slaughter of the helpless animals, killing one hundred and fifty of them, for which they had no use at all. It is certain that so able a general as Crazy Horse had scouts watching Crook all the time, and would have discovered his advance in any event; but with all the noise made by the Indians in the buffalo hunt, there was no possibility of a surprise. As a matter of fact, it was Crazy Horse who began the game. Crook was ready for him.

V. The Battle of the Rosebud

About half after eight o'clock in the morning, the resting soldiers were called to attention by the sound of shots from the bluffs in front of them, over which their allies had disappeared. It was at first supposed that these friendly Indians had run across another herd of buffalo, but a few moments told the practised troopers that the firing was the beginning of a battle rather than that of a hunt. At the same time the Indian auxiliaries came galloping back to the main body at full speed, yelling:

"Sioux, Sioux! Heap Sioux!"

Without waiting for orders, the troopers saddled their horses and fell in. They got ready none too soon, for right on the heels of the fleeing Crows and Shoshones came the Sioux. In front of them to the right, the left, the low bluffs inclosing the plain, were ringed with Indians in full war-gear. As one observer described it to me, they looked like swarms of blackbirds, there were so many of them and in such rapid motion. They kept coming and coming into view, and as they dashed up to the brink of the hills upon their war ponies they opened a long-range fire upon the soldiers, which from the distance did little damage. There were at least a thousand of them in plain sight. How many others there might be, no one could tell. It was a safe guess that those in sight constituted but a small part of the force.

It is said that there were at least six thousand warriors that day under the command of Crazy Horse, but that most of them were not engaged. Crazy Horse had planned an ambush for General Crook, and he had hoped to defeat him by luring the soldiers into it, or by separating the army into small detachments and overwhelming them in detail. His plans were well devised, and came very near being successful. That they did not succeed is probably due more to the acts of the Indians themselves than because of the wariness of the soldiers.

Crook acted at once. Sending his staff officers to rally the Crows and Shoshones, he directed them to circle to the right and left, and make ready to fall on the flanks of the Indians. Mills, who had behaved so gal-

Courtesy of The Century Co.

THE DIFFICULT TASK OF THE HORSE-HOLDERS IN ACTION

Drawing by Frederic Remington

lantly at the Tongue River, was ordered to charge the Indians straight up the valley to the bluff to the northward, the front. Two troops of Van Vliet's squadron were rushed off to the southward, the rear, to seize a commanding position to prevent the Indians from circling around in that direction and getting in Crook's rear. The infantry and part of the Second Cavalry were dismounted, and thrown forward as skirmishers around the foot of the bluffs. Royall took Henry's battalion, with Van Vliet's remaining troop, one of Mills' troops which he detached while the battalion was on the gallop, and another of Noyes' troops, and charged the Indians on the left.

Mills' charge was most gallantly delivered. The soldiers struggled through the bog, raced across the bottom land for about eight hundred yards, and scrambled up the bluffs in twenty minutes, finding themselves, when they reached the top, within fifty paces of the Sioux. There was no time to use carbines. Firing revolvers, the battalion rushed at the Indians. The savages fired ineffectively, gave way, and fled instantly to higher ground six hundred yards further on, where they opened fire. In their excitement they shot badly. Mills dismounted his battalion, deployed them as skirmishers, rushed the second ridge and cleared it, the Indians sullenly retiring before him, and again opened fire on the troops, to which the cavalrymen made effective reply. The Sioux galloped rapidly to and fro, yelling and firing from their horses, kicking up clouds of dust, but doing little harm.

Royall, Henry, and Van Vliet had a similar success on the left, where the ground was much more open and unfavorable for defense, although the Indians were massed more heavily in that quarter than before Mills.

Meanwhile the Crows and Shoshones had fallen upon the flanks of the Sioux, but not very effectively. Every one in the field except a small reserve was now hotly engaged. The pressure on Mills became stronger, but he drove the Indians from him by another gallant attack. Thereafter he was reinforced by Noyes' battalion. The front of his line was finally partially cleared by this last dash. The Indians who had been attacking him thereupon left him, and joined the others before Royall and Henry.

Crook now withdrew Mills' command from the battlefield, and Mills was ordered to take his three troops down the Dead Cañon of the Rosebud and attack the villages which it was believed the Indians were defending. Mills' movements were supported by the five troops of the Second Cavalry under Noyes. Crook promised to follow up the movement, and support it with the remaining cavalry and infantry. We will follow this movement later.

Mills' place in the line was occupied by Tom Moore and his packers and some other auxiliaries from the camp, and a smart fire was kept up in that direction. On the left the firing was fast and furious. The Indians from the front cleared by Mills joined their associates on the left, and again and again attacked Royall, Henry, and Van Vliet, who had joined the other two, with the most determined courage. Charge and counter charge were made over that portion of the field. Now the troops gave way before the Indian advance, now the soldiers were rallied and hurled back the Indians, now the Indians retreated before some desperate countercharge. So went the varying fortunes of the hour. The number of savages increased with every passing moment. To the eyes of the astonished soldiers they seemed to spring from the ground. If one fell in the line, a dozen were ready to take his place.

In one of the charges Captain Henry was shot through the face and frightfully wounded. The troopers had dismounted, but the officers remained mounted. Henry

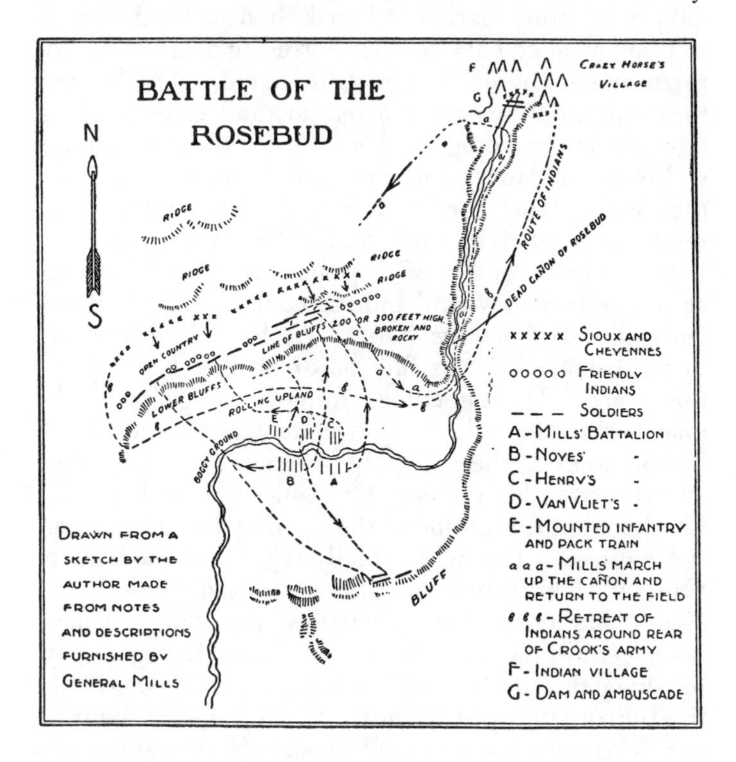

reeled in his saddle as the bullet pierced both his cheek bones and tore out the whole front of his face below the eyes. Although, as an eye-witness has it, he was spitting blood by the handful, he continued on the battle line. The situation of Royall's wing of Crook's army was precarious. Henry's battalion held the extreme left flank. It was his duty to remain there. Vroom's troop L, of the Third Cavalry, had become separated from the main body during the battle, and was caught ahead of the line and surrounded by Indians, in imminent danger of annihiliation. Crook had ordered Royall to bring his men back to their horses, in order to mount them and prepare for a general charge. The Indians took this movement for a retreat, and came dashing after the retrograding troopers. Only the cool courage of Royall and Henry, and the magnificent way in which they handled their forces when they went forward to the rescue, prevented the annihilation of Vroom's troops.

It was in the midst of this operation that Henry received his fearful wound and stayed on the line.* Presently he fell from his horse. As he did so, the soldiers, dismayed by his fall, began to give back before the Sioux. The impetuous Indians charged over the place where Henry lay. Fortunately, he was not struck by the hoofs of the galloping horses. His men rallied and rushed to his rescue. Old Chief Washakie and his Shoshones at this period of the fighting displayed splendid courage. The fight actually raged over the body of Captain Henry until the Indians were driven off, and Henry was rescued from what would otherwise have been certain death. After this fierce struggle, part breaking through the line and part turning the flank, the Indians galloped down the valley between the river and the troops, and finally disappeared on the other side of the Dead Cañon, their retreat accelerated by the movement of Mills toward the villiage. The fighting had lasted a little more than two hours.

* For a sketch of General Henry and an account of his experiences in this battle and elsewhere, see the last chapter of this book.

VI. Mills' Advance down the Cañon

Meanwhile Mills and his men, in a column of twos, trotted down the gloomy depths of the Dead Cañon, the rocky walls of which, towering on either side, would have afforded abundant cover for Indian riflemen. Before entering the cañon, they had cleared the mouth of it of a body of Sioux by a smart charge, and they were thereafter unmolested. They advanced rapidly but with caution, although what they could have done if attacked it is hard to see, and how caution would have saved them it is difficult to tell. They had their orders to go through the cañon and attack the village. There was nothing to do but obey. Sending them forward was a mistake which might have resulted in a terrible disaster, although nobody believed that then. The soldiers had not yet realized what fighters these Indians were. The Custer disaster was still to come, and no one imagined that so large a body as that commanded by Mills and Noyes could be defeated. If Crook had followed with his whole force, the troops under his command would have been annihilated; it is probable that not one of them would have come out of that cañon.

When Crook began to prepare to follow Mills with the rest of his force, he discovered that he had a much larger number of wounded than he had thought possible, and the doctors protested against their being left with a feeble guard while Crook with the best of the force went up the cañon. The protest was justified by the situation. Besides, the attack on Royall and Henry had not yet ceased. Crook reflected, concluded that he could not leave the field, and that Mills' force was too weak for the work assigned it. The general thereupon depatched

Captain Nickerson, of his staff, attended by a single orderly, at the imminent peril of their lives, with orders to ride after Mills and tell him to leave the cañon, defile to the left, and rejoin him at once. Crook hoped that Mills, on his return to the field, might succeed in getting in the rear of any Indians who might be lurking in the hills before Royall's shattered line.

So rapid had been Mills' movements, that Nickerson, although going at the full speed of his horse, did not overtake him until he had penetrated some seven miles down the cañon. Fortunately for all concerned, the command had halted where a cross cañon made an opening toward the west, and on that side the cañon was so broken and so sloping that it could be scaled by the troopers. Firing was heard to the front, and the Indians were detected massing to attack Mills' detachment. A halt had been ordered for the purpose of making final preparations for the attack.

"Mills," said Nickerson, as he came galloping up, "Royall has been badly handled — there are many wounded. Henry is severely hurt, and Vroom's troop is all cut up. The General orders that you and Noyes defile by your left flank out of this cañon and return to the field at once. He cannot move out to support you and the rest on account of the wounded."

Never was order more unwelcome. The officers at the head of the column urged Mills to go on. The Indian village was in sight. Crook could not have known how near they were, or he would not have recalled them. Mills, however, was a thorough soldier. In his mind orders were to be obeyed, and he silenced the objectors and advisers, and did as he was directed, although with great disappointment and reluctance. Never was obedience better justified. General Mills admits now that, had he disobeyed Crook, his command would have been annihilated.

The cañon was the mouth of Crazy Horse's trap. A short distance farther on, it ended in a great dam covered with broken logs, making a dangerous abattis. Here the main body of the Indians had been massed. Here they expected, seeing the confident advance of the eight troops of cavalry up the cañon, to fall upon them and kill them all, which they might easily have done. Nickerson got there just in time.

Mills instantly turned to the left and led his troops up the broken wall of the cañon to the high ground on the farther side; fortunately, he had been overtaken at about the very point where the ascent was practicable for troops. Presently the detachment rejoined the main body, their progress being unmolested.

There were ten soldiers killed and twenty-seven seriously wounded, besides a great number of slightly wounded. Most of the casualties were in Royall's command, Vroom's troop having lost heavily while it was in such peril.

Crook camped for the night on the battle-field. The dead were buried, the wounded looked after temporarily, and the next morning the soldiers withdrew. They went back to their camp at Goose Creek and stayed there. The battle was in one sense a victory for the white soldiers, in that they drove the Indians from the field, forcing them back at least five miles. In another, and a larger and more definite sense, it was a decided victory for Crazy Horse. He had fought Crook to a standstill. He had forced him back to his base of supplies. He had stopped the farther progress of that expedition. He had protected his villages and had withdrawn his army in good order.

If Mills' command had not been recalled, it is certain that it would have been annihilated. As it was, the Indians had done remarkably well. Crazy Horse, free from further apprehension of pursuit by Crook for the present, had leisure to turn his attention to the other two expeditions, which there is no doubt he was well aware had been launched against him.

While technically it was perhaps a drawn battle, as a feat of arms the battle of the Rosebud must go down to the credit of the Indians. It was more like a pitched battle than any that had been fought west of the Missouri heretofore. The individual officers and soldiers of the army did splendidly; so did the Indians. Mills had displayed commendable dash and daring in all his charges. Royall, Henry and Van Vliet, and Chambers with the infantry, had fought skilfully and bravely against an overwhelming force. Crook's dispositions were good on the field, and were well carried out by his subordinates. The same may be said of Crazy Horse, his subchiefs, and their warriors.

Crook had nearly exhausted his ammunition in the hard fighting, the larger part of his supplies had been expended, and he had a number of very seriously wounded on his hands. There was not one chance in a thousand that he could catch the Indians now. There was nothing left for him to do but go back to the main camp, send his wounded back to Fort Fetterman for treatment, order up more supplies and more troops, and await a favorable opportunity to attack again.

To anticipate events, it may be noted that, owing to the disaster to Terry's column, Crook did not advance until August.

CHAPTER TWO

Ex-Trooper Towne on the Rosebud Fight

I AM afraid that any attempt on my part to comply with your request will be a very feeble attempt to describe to you the Battle of the Rosebud, which took place on June 17th, 1876. There are many men living who participated in that battle who can describe more fully and more comprehensively than I the details of that day. However, I will do my best.

On the 16th day of June, 1876, General Crook with his command was camped on the Tongue River awaiting the arrival of three hundred Crow and Shoshone Indians to be used as scouts, under Frank Gruard, a noted scout of the Indian country, it being Crook's intention thoroughly to scout the whole country from the Powder and Tongue Rivers north to the Yellowstone, and to co-operate with the other columns in the field under Custer, Terry, and Gibbon.

At about five P.M. on the afternoon of June 16th the three hundred scouts came into our camp, and shortly afterward General Crook gave orders to the command to prepare for a night march. Extra ammunition and extra rations were issued, and at about eight P.M. we broke camp and mounted into the saddle to commence our march into the Indian country, which was overrun by the Oglalas, Brulés, Unkpapas, and Miniconjous, the four most powerful tribes of Sioux Indians on the plains, for it is to be remembered that the whole Sioux Nation had left their reservations and was then on the war-path.

General Crook had, on the morning of the 16th, sent out scouts to find and report any Indian signs that might be found. Numerous signs were found which indicated that a large party of Indians had recently passed that way going in a northerly direction, with the evident intention of joining those from the Brulé agency on the Yellowstone. It was General Crook's purpose to cut them off. Thus the forced march of the night of the 16th.*

After a long and tiresome journey of all night, about seven A.M. of the 17th Indians were seen on the hills to our front and left who were evidently watching our movements. It was reported to General Crook by the scouts that we had gotten into a country that was completely alive with hostile Indians and that we were near an immense Indian camp.

General Crook at about 7.30 A.M. went into camp with the intention of making another night march and, if possible, overtake their camp the next day. As we had been in the saddle all night, men and horses needed a few hours rest.

After going into camp we unsaddled and put our horses to graze, but first hobbling them to prevent any stampede that might be attempted. While we were putting our horses to graze the whole range of hills in our front became literally alive with Indians, and at the

* Trooper Towne is in error here ; there was no night march, according to Captain Bourke. See his "On the Border with Crook." See also "War Path and Bivouac," by Finerty.—C. T. B.

same time the Crow and Shoshone Indians with us commenced their warlike preparations by daubing themselves with war paint and riding their ponies in a circle one behind the other, and at the same time singing their war songs.

After riding in this manner a short while, the circle broke and the whole group of Indian scouts charged up the hill toward their enemies. It is a well known fact that the Sioux and Crow Indians were enemies toward each other at that time. I have seen the Crow Indians shoot buffalo and let them lie where they fell, not even undertaking to remove the hide, because, they would say, "Sioux Buffalo no good," which indicated that as the Sioux Indians were their enemies, so were the buffalo found in the Sioux country their enemies also. Everything in the Sioux country was an enemy to the Crow Indian.

While our Indians were making their charge upon the Sioux, General Crook gave orders to saddle up, for well he knew that a battle was on hand. After we had saddled and formed in line, my troop, F, Third Cavalry,* was placed on the left flank of the command, and it with two other troops were detailed as skirmishers and were ordered to make a flank movement to our left and gain the hills, where we dismounted, leaving each fourth trooper to hold the horses. We then formed the skirmish line on foot, which was commanded by Lieutenant-Colonel Royall.

At this time I witnessed a most daring act by a bugler by the name of Snow, who was carrying a despatch from General Crook to Colonel Royall. General Crook was stationed on one of the hills to our right, near the center

*This troop was commanded by Lieutenant Reynolds, and was in Henry's battalion.—C. T. B.

of the line,* where he could view all that was taking place. Wishing to send an order to Colonel Royall, he directed his orderly, Bugler Snow, to carry it with all haste. The most direct route was down a steep hill and across a level plain and then up another hill, where Colonel Royall was. All chances of reaching there alive were against him.

When I saw him he was coming as fast as his horse could carry him, while two Indians were after him with the intention of capturing him. Seeing that they could not capture him, they finished the game by shooting at him, and proved their good marksmanship as poor Snow fell from his horse, shot through both arms, but he delivered his orders all right.

After remaining on the skirmish line for perhaps two hours, we were ordered to fall back and remount our horses to take a new position (our horses were held in check in a ravine), as it was impossible to hold our present position against such overwhelming odds. I must say that I never saw so great a body of Indians in one place as I saw at that time, and I have seen a great many Indians in my time. It seemed that if one Indian was shot five were there to take his place. If we had remained in our first position we would all have been killed, and I consider that we retreated in the right time.

I had not gone more than one third of the distance from our position to where the horses were when I overtook three other soldiers of my own troop carrying a sergeant by the name of Marshall, who had been shot through the face. I knew that time was precious and none to lose. I could not give them the cold shoulder by passing them without giving a helping hand. Glanc-

* Crook was right in the fighting ; his horse was shot under him.—C. T. B.

ing back, I saw the hostiles coming over the hill. I said to the others, "Quick, here they come!"

At that instant my comrades, to save themselves, dropped the wounded sergeant and hastened to their horses. The sergeant, seeing that I was the only one left, said:

"Save yourself if you can, because I am dying. Don't stay with me." I replied:

"Dave, old boy, I am going to stay right here with you and will not desert you."

Grasping him with all my strength, I carried my comrade until it was useless to carry him any farther, for he was dead. I then laid him down and left him and hurried to get away.

I don't think that I had gone more than ten yards when I was surrounded by about twenty or more of the most murderous looking Indians I ever saw. You can talk of seeing devils; here they were in full form, painted in the most terrifying manner, some with their war bonnets adorned with horns of steers and buffalo. It was enough to strike terror to anyone's heart.

I knew that my time had come, I knew that I would be taken prisoner. I fought, but it was fighting against terrible odds. There I was down in that ravine, alone and in the midst of a lot of murderous savages.

Taking my carbine from me and throwing a lariat over my head and tightening it about my feet, I was helpless. This was all done in an instant, while I struggled and fought in vain, until I was struck on the head with something which rendered me unconscious and caused me to fall. As I went down a bullet struck me in the body.

I think that when the bullet struck me I regained my consciousness, because I realized I was being dragged at a lively pace over the ground by a pony at the other end of the lariat. It was, I think, the intention of the Indians either to drag me to death at the heels of the pony or after getting me away to torture me in some other manner.

They captured one other comrade of mine by the name of Bennett, of L Troop, Third Cavalry, and completely cut him in pieces. His remains were buried in a grain sack.

After I was dragged in this manner for some distance, my captors were charged by one of the troops of cavalry, and to save themselves from capture abandoned me and made their escape. Thus was I enabled to regain my liberty.

I was immediately sent to the field hospital, and three days later I, with eighteen other wounded men, was sent to the post hospital at Fort Fetterman. You ask in your letter did I get a medal of honor for trying to get my sergeant away. I am very sorry to say that I did not, although I do think that even at this late day had I some one who would speak a good word in my behalf I think that my case would be taken up and that I might get one.*

I receive a very small pension for the wound received in this Indian battle, and that is all my recompense.

Hoping that this narrative of my experience in the Battle of the Rosebud may be of interest, I have the honor to remain,

Yours sincerely,

PHINEAS TOWNE.

St. Louis, Mo.

* The official records show that Marshall was killed and Towne wounded in the battle. If this account falls under the eye of any one in authority, I trust an investigation may be made, and that the medal may be awarded, if it has been earned.—C. T. B.

CHAPTER THREE

The Grievance of Rain-in-the-Face

I. The Yellowstone Expedition of 1873

HAVING thus disposed of the most formidable column, Crook's, in so summary and so effective a manner, the Indians under their able leadership turned their attention to Custer and Gibbon.

Before the Little Big Horn campaign is discussed, however, in order the better to understand the most terribly dramatic episode in the most disastrous of our Indian battles, it will be necessary to go back a little and take up the thread of the discourse later.

The country watered by the Yellowstone and its affluents, traversed by the Black Hills and other ranges of mountains, and protected by the almost impassable Bad Lands in Dakota, had been up to 1873 practically a terra incognita. However, the Northern Pacific Railroad was even then surveying a route across it. Gold had been discovered, and miners and settlers were crowding in. The Indians, since the treaty of 1868, which had resulted in the abandonment of Fort Phil Kearney and the other posts, had been ugly in mood and troublesome in action. They welcomed neither railroad nor men.

An expedition of some seventeen hundred men under General Stanley was sent into the country in 1873. Custer and the Seventh Cavalry formed a large part of the command. There were no guides. The country, especially in the Bad Lands, was a terrible one to cross, and Custer volunteered to take two troops of cavalry and ride some miles ahead of the main body every day to mark a road. Custer possessed a faculty for this sort of work which was simply marvelous. He was a born pathfinder, better even than Frémont.

On the 4th of August he left camp at five o'clock in the morning with the troops of Moylan and Tom Custer, eighty-six men, five officers, and a favorite Arikara scout, called Bloody Knife. At ten A.M., reaching the crest of some bluffs along the river bank, they saw spread before them a beautiful village, through which the river gently meandered between the tree-clad banks. They advanced two miles up the valley, and made camp under the trees for a noonday rest. They had come at a smart pace and were far ahead of the main column, which was out of sight in the rear. The passage through the valley was easy, and there was no necessity for them to press on. The weather was hot. After picketing the horses, partaking of their noonday meal and posting sentries, officers and men threw themselves on the grass and fell asleep.

At one o'clock the sentry on the edge of the timber gave the alarm. A small party of Indians was approaching, in the hope of stampeding the horses. All Indian attacks begin that way. After the horses are stampeded the soldiers have to fight where they are, and, as the Indians are mounted, the dismounted troopers are at a disadvantage. Custer was on his feet in an instant, shouting:

"Run to your horses, men!"

The troopers were no less alert. Before the Indians could stampede the horses, each man had reached his animal and led him back into the timber. A few shots drove off the little party of savages, the horses were saddled, and the men moved out. As they did so, six mounted Indians appeared on the crest of a little hill. Custer led the way toward them. They retreated slowly, keeping just out of range. In this manner they drew the soldiers some two miles up the valley.

Finally, in the hope of getting near to them, Custer took twenty men, with his brother and Lieutenant Varnum in command, and rode out some two hundred yards ahead of the remainder under Captain Moylan, who were directed to keep that distance in rear of the advance. Custer, accompanied only by an orderly, rode about the same distance ahead of the advance, making peace signs to the six Indians whom they had pursued. As he approached nearer to them, their pace slackened and they suddenly stopped.

To the left of the soldiers was a thick wood. It occurred to Custer that Indians might be concealed therein, and he sent his orderly back to the advance to caution them to be on their guard. Scarcely had the orderly reached the advance when the Indians they had been pursuing turned and came at full gallop toward Custer, now alone in the valley. At the same instant, with a terrific war whoop, three or four hundred splendidly mounted Sioux burst out from the trees on the left.

Custer was riding a magnificent thoroughbred. In a second he was racing for his life toward the advance-guard. The Indians had two objects in view. They wanted to intercept Custer and also cut off the advance party from Moylan's men, who were coming up at a gallop. Only the speed of Custer's horse saved his life. As he galloped toward them, he shouted to Tom Custer to dismount his men. He was not heard in the confusion, but young Custer knew exactly what to do. While five men held the horses, the other fifteen threw themselves on the ground. On came the Indians after Custer. As soon as they were within easy range, the dismounted men blazed away right in their faces. The troopers were armed with breech-loaders, and the first volley was succeeded by a second. Several of the savages were hit and many of their horses. They reeled, swerved, and Custer rejoined his men. A few moments after, Moylan came up with the main body.

Custer now dismounted most of his men, and keeping a bold front to the Indians, retreated in the timber, fighting hard all the way. Reaching the river, they made good their defense. The Indians tried all their devices to get them out. They set fire to the grass, but it was green and did not burn readily. All their efforts to dislodge the troopers failed, and late in the afternoon a heavy squadron came up on the gallop from the main body under Stanley and put them to flight. It was a sharp affair, and the Indians suffered severely.

The only losses to the expedition on that day were two civilians: Doctor Honzinger, a fat old German, who was the veterinarian of the regiment, and Mr. Baliran, the sutler. They were both quiet, inoffensive, peaceable men, very much liked, especially the doctor. They were amateur naturalists, and frequently wandered away from the main body on botanizing excursions. They had done so that morning, following Custer's advance, and the Indians had fallen upon them and murdered them. It was the discovery of the remains of these two men which had caused General Stanley to despatch the cavalry to the relief of the advance.

Bitter was the anger of the officers and men over this murder of unarmed non-combatants, and deep and abiding was their thirst for vengeance on the Indians who perpetrated it.

II. The Capture of Rain-in-the-Face

The next year, 1874, the Seventh Cavalry being stationed at Fort Abraham Lincoln, near Bismarck, on the upper Missouri, word was brought to Custer by a scout that a famous Sioux, called by the picturesque name of Rain-in-the-Face, was at Standing Rock Agency, some twenty miles away, boasting that he had killed Doctor Honzinger and Mr. Baliran. Rain-in-the-Face was already a renowned warrior, of more than ordinary courage. That he should have left the hostiles under Sitting Bull to come to the agency was a thing implying peculiar bravery; and that he should there openly boast of the murder was even more extraordinary. Custer immediately determined upon his capture, although to effect it would be a matter of difficulty and danger.

The agency was filled with Indians waiting for the issue of rations; and, though they were on a peaceable errand, they were always unruly, insubordinate, and on the alert. Captain Yates and Captain Tom Custer, with one hundred troopers, were detailed to make the arrest. The arrival of one hundred men at the agency instantly excited the suspicion of the Indians. To divert it from the real object, Captain Yates ostentatiously detached a lieutenant with fifty men to ride to some villages ten miles away in quest of certain Indians who had some time before raided a settlement and run off some stock, killing the herders. With the remainder he purposed to wait for the return of the detachment. Meanwhile it was learned from a scout that Rain-in-the-Face was in the sutler's store.

Tom Custer, with five picked men, was ordered to enter the store and make the arrest. The store was full of Indians. The weather was very cold, and the Indians kept their blankets well around their faces. It was impossible to tell one from another. Tom Custer had received a good description of Rain-in-the-Face, however, but it availed him nothing under the circumstances. He and his men, therefore, mingled freely with the Indians from time to time, making small purchases of the sutler to divert suspicion as they lounged about the store. They deceived the savages entirely, in spite of their watchful scrutiny and suspicion. At last one Indian dropped his blanket and stepped to the counter, either to speak to the trader or to make a purchase.

It was Rain-in-the-Face. Custer recognized him immediately. Stepping behind him, he threw his arms about him and seized him in an iron grasp. The Indian, who had observed the movement too late, attempted to fire his Winchester; but Custer was too quick for him. The five troopers sprang to the side of their captain, disarmed Rain-in-the-Face, and presented their guns to the astonished and infuriated Indians. The room was filled with seething excitement in a moment. The Indians surged toward the troopers, and perhaps would have made short work of them, had not Captain Yates at this juncture entered the room with a detail of his men.

Rain-in-the-Face, a magnificent specimen of Indian manhood, had ceased to struggle the moment he was convinced that it was unavailing. He was led outside,

THE CAPTURE OF RAIN-IN-THE-FACE

Drawing by E. W. Deming

securely bound and mounted on a horse. The troopers were assembled, and in spite of threats and menaces by the Indians, who did not venture to attack, they started back to Fort Lincoln with their prisoner.

Messengers were sped in every direction to the different bands of Indians to mass a force to release Rain-in-the-Face, who was a man of such importance, being the brother of Iron Horse, one of the principal chiefs of the Unkpapas, that no price was counted too great to secure his liberty; indeed, before starting, they had offered Yates two warriors in exchange for him. The rapidity with which the troops moved was such that the prisoner was safely imprisoned at Fort Lincoln before anything could be done.

Rain in-the-Face stubbornly refused to say anything for a day or two, but finally made full confession that he had shot Mr. Baliran and wounded Doctor Honzinger, who had fallen from his horse, whereupon he had crushed his head with stones. He was put in the guard-house preparatory to being tried for murder, and kept there in spite of the efforts to release him that were made by many prominent Indians. In the same guard-house were some civilians who had been caught stealing grain. One bitterly cold night, during a raging blizzard, the civilians, with some outside assistance, succeeded in making their escape. Rain-in-the-Face took advantage of the opportunity and left also. He joined the hostiles under Sitting Bull, and sent back word that he intended to have his revenge on the Custers for the treatment he had received.

CHAPTER FOUR

The Little Big Horn Campaign

I. Custer Loses His Command

TO return to the spring of 1876. When the column which Custer was to have commanded moved out, Custer led his own regiment, while Major-General Alfred H. Terry was in personal command of the column. I give the reason in the words of General George A. Forsyth in a recent letter to me:

"For some reason Custer, one of the most splendid soldiers that ever lived, hated General Belknap, the Secretary of War. He was a good hater, too. When General Belknap was imprisoned and undergoing trial Custer wrote that he knew of certain things regarding the appointment of post-traders on the upper Missouri River, which things the prosecution thought were what they needed to insure conviction. As a matter of fact, Custer did not know anything. He had heard disappointed men who had failed to get said post-traderships curse Belknap and say that they knew Belknap had sold the traderships to the appointees. It was not so. Belknap had given these appointments to certain able Iowa politicians for their friends, in order to secure their influence in the next campaign for United States Senator from Iowa, as he had determined to try for a senatorship from his state, viz., Iowa.

"It was entirely within his own right to make these appointments and there was really nothing wrong in doing so. Of course the disappointed applicants were furious, and especially certain men who had served with Belknap during the Civil War and who thought they had a claim on him. They could not tell lies fast enough about Belknap and especially to Custer, who was thoroughly honest and believed what they said. This was what Custer thought he knew.

"Custer was summoned to Washington of course. When he was questioned by the House Committee of prosecution it was apparent that he did not know anything. His evidence was all hearsay and not worth a tinker's dam. The President — General Grant — was indignant at Custer's statements regarding Belknap, which turned out to be all hearsay. . . . The President directed General Sherman not to permit Custer to take the field against Sitting Bull — undoubtedly to punish him.

"You will recall that Belknap was — in a sort of Scotch verdict way, 'Not proven, my lord' — acquitted. It was only upon the strong, insistent and urgent request of General Sheridan to General Sherman — the then Commanding General of the Army — that the President finally said that if General Sheridan regarded Custer's services of great importance in the campaign, Sherman might authorize Sheridan to permit him to join his regiment and serve under General Terry, who was appointed to command the expedition. Sherman wired Sheridan what the President said, and Sheridan at once applied for Custer as in his opinion 'necessary.'*

* It was General Terry's urgent representations which were the main-springs of Sheridan's action. — C. T. B.

"I was in Europe at the time of the Custer disaster, and on my return to General Sheridan's headquarters I saw all the correspondence in the case."

Therefore, instead of commanding the column, Custer was placed under Terry, who was to command Gibbon's column as well, when the junction had been made between the two. On the 17th of May the command left Fort Lincoln. The seriousness of the situation was felt as never before in an Indian campaign. It was realized that no child's play was before the troops, and it was with unusual gravity that the regiment marched away. Mrs. Custer tells how General Terry ordered the force to parade through Fort Lincoln to reassure the women and children left behind by the sight of its formidable appearance.

The best part of the expedition was the Seventh Cavalry, six hundred strong, with Custer at its head. The band played "Garry Owen," the famous battle tune of the Washita, as they marched away. They halted on the prairie afterward, and an opportunity was given to the officers and men to say good-by to the dear ones to be left behind; then, to the music of "The Girl I Left Behind Me," they started on that campaign from which half of them never came back.

They reached the Powder River without mishap, and were there joined by General Gibbon, who reported his command encamped along the Yellowstone, near the mouth of the Big Horn. Major Reno, of the Seventh Cavalry, with six troops had been sent on a scouting expedition to the southward, and had discovered a big Indian trail leading westward toward the Big Horn country. On the 17th of June Reno's men had been within forty miles of the place where Crook was fighting his fierce battle, although, of course, they knew nothing of it at this time. On the 22d Custer was ordered to take his regiment with fifteen days' rations and march down the Rosebud, thoroughly examining the country en route until he struck the Indian trail reported by Reno.

II. Did Custer Obey His Orders?

And now we come to the most important question of this remarkable campaign. On the one hand, General Terry has been severely censured for its dire failure; the death of Custer and the escape of the Indians have been laid at his door. On the other hand, it has been urged that Custer disobeyed his orders, broke up Terry's plan of campaign, and by his insubordination brought about a terrible disaster and let slip the opportunity for administering a crushing defeat to the Indians, which probably would have ended the war and prevented a deplorable loss of life, to say nothing of prestige and treasure. Both officers had, and still have, their partizans, and the matter has been thoroughly threshed out.

As between Custer and Terry, I profess absolute impartiality, although, if I have any natural bias, it is toward Custer, whose previous career, as I have investigated it, appeals to me more than Terry's, distinguished as were the latter's services. I have studied the situation carefully, examining all the evidence published by both sides, and very reluctantly, in spite of my liking for poor Custer, I am compelled to admit that he did disobey his orders; that his action did break up a most promising plan, which, it is highly probable, would have resulted in a decisive battle with the Indians and the termination of the war; and that he, and he alone, must be held responsible for the subsequent disaster.

General Terry's order to Custer, which follows, is entirely clear and explicit:

Camp at Mouth of Rosebud River, M. T.,
June 22d, 1876.

Lieutenant-Colonel Custer, 7th Cavalry. Colonel:

The Brigadier-General Commanding directs that, as soon as your regiment can be made ready for the march, you will proceed up the Rosebud in pursuit of the Indians whose trail was discovered by Major Reno a few days since. It is, of course, impossible to give you any definite instructions in regard to this movement, and were it not impossible to do so, the Department Commander places too much confidence in your zeal, energy, and ability to wish to impose upon you precise orders which might hamper your action when nearly in contact with the enemy. He will, however, indicate to you his own views of what your action should be, and he desires that you should conform to them, unless you should see sufficient reasons for departing from them. He thinks that you should proceed up the Rosebud until you ascertain definitely the direction in which the trail above spoken of leads. Should it be found (as it appears almost certain that it will be found) to turn toward the Little Horn,* he thinks that you should still proceed southward, perhaps as far as the head-waters of the Tongue, and then turn toward the Little Horn, feeling constantly, however, to your left, so as to preclude the possibility of the escape of the Indians to the south or southeast by passing around your left flank. The col-

* At the time this was written, it was not generally understood that the full Indian appellation of this stream was Little Big Horn. — C. T. B.

Courtesy of The Century Co.

GEN. GEORGE ARMSTRONG CUSTER

Killed with half his regiment at the Little Big Horn

umn of Colonel Gibbon is now in motion for the mouth of the Big Horn. As soon as it reaches that point it will cross the Yellowstone and move up at least as far as the forks of the Big and Little Horns. Of course its future movements must be controlled by circumstances, as they arise, but it is hoped that the Indians, if upon the Little Horn, may be so nearly inclosed by the two columns that their escape will be impossible.

The Department Commander desires that on your way up the Rosebud you should thoroughly examine the upper part of Tullock's Creek, and that you should endeavor to send a scout through to Colonel Gibbon's column, with information of the result of your examination. The lower part of this Creek will be examined by a detachment from Colonel Gibbon's command. The supply steamer will be pushed up the Big Horn as far as the forks if the river is found to be navigable for that distance, and the Department Commander, who will accompany the command of Colonel Gibbon, desires you to report to him there not later than the expiration of the time for which your troops are rationed, unless in the meantime you receive further orders.

Very respectfully your obedient servant,

E. W. SMITH,
Captain 18th Infantry,
Acting Assistant Adjutant General.

Custer was directed to march southward until he struck the trail Reno had discovered. If, as Terry supposed, it led across the Rosebud, he was not to follow it westward to the Little Big Horn, or until he met the Indians, but he was to turn to the southward until he struck the head-waters of the Tongue River. If he found no Indians there, he was to swing northward down the valley of the Little Big Horn, toward the spot where Terry supposed the Indians to be, and where, in reality, they were. Meanwhile Gibbon was to come up the Little Big Horn from the north toward the same spot. In the general plan of the campaign, Crook and his force were supposed to prevent the Indians from moving south — which they did, by the way. Custer was to keep them from going east, and, as he advanced, was "to feel to his left" to preclude all possibility of their slipping between him and Crook, while Gibbon was to keep them from going off to the north. The Indians would have no direction open to them for flight except westward, and in that case the troops hoped to overtake them in a difficult country, inclosed by mountains and rivers.

Terry, although he was not an experienced Indian fighter, had divined the position of the Indians with remarkable accuracy, and he fully expected to find them on the Little Big Horn. If Custer had followed Terry's orders, he would have reached the Indians on the day that Gibbon's men, as we shall see, rescued Reno. After the disaster Terry magnanimously strove at first to conceal from the public the fact that Custer had disobeyed his orders. Custer had paid the penalty for his disobedience with his life, and Terry was willing to bear the odium of the defeat and failure. His self-sacrifice was noble and characteristic; but a mistake, caused by the carelessness of General Sherman, coupled with the enterprise of a brilliant newspaper reporter, who posed as a regularly accredited government messenger, defeated Terry's intent, and instead of the first report, which made no allusion to the disobedience of orders, being made public,* a second report, which told the whole

* It was delayed in transmission, owing to the cutting of the telegraph wires by the Indians.

story, and which was intended for the authorities alone, was given to the press and immediately spread broadcast. The first report soon turned up, and Terry thereafter was made the victim of unmerited obloquy by Custer's partizans, who said that the absence of any mention in the original report of any disobedience on the part of Custer, and the alleged failure to allude to the plan of campaign which Custer had frustrated, was evidence that no importance was attached to the plan by Terry or any one until after the failure and consequent popular indignation. Terry's answer to this was a noble silence, to save Custer's reputation. The living assumed the responsibility to protect the fame of the dead — honor to him!

General Gibbon also has gone on record in a letter to Terry regarding the situation:

"So great was my fear that Custer's zeal would carry him forward too rapidly, that the last thing I said to him when bidding him good-by after his regiment had filed past you when starting on his march was, 'Now, Custer, don't be greedy, but wait for us.' He replied gaily as, with a wave of his hand, he dashed off to follow his regiment, 'No, I will not.' Poor fellow! Knowing what we do now, and what an effect a fresh Indian trail seemed to have had upon him, *perhaps we were expecting too much to anticipate a forbearance on his part which would have rendered coöperation of the two columns practicable.**

"Except so far as to draw profit from past experience, it is perhaps useless to speculate as to what would have been the result had your plan, as originally agreed upon, been carried out. But I cannot help reflecting that in

* Italics mine.—C. T. B.

that case my column, supposing the Indian camp to have remained where it was when Custer struck it, would have been the first to reach it; that with our infantry and Gatling guns we should have been able to take care of ourselves, even though numbering only about two-thirds of Custer's force; and that with six hundred cavalry in the neighborhood, led as only Custer could lead it, the result to the Indians would have been very different from what it was."

With regard to Gibbon's generous suggestion that Custer was suddenly carried away by the opportunity presented, the testimony of the late General Ludlow is interesting. According to him, Custer stated on the 8th of May, in St. Paul, Minnesota, that he intended, at the first chance he got in the campaign, to "cut loose from (and make his operations independent of) General Terry during the summer;" that he had "got away from Stanley and would be able to swing clear of Terry." *

It is difficult, nay, it is impossible, therefore, to acquit Custer of a deliberate purpose to campaign on his own account so soon as he could get away from General Terry. The sentence of Terry's orders commencing: "It is, of course, impossible to give you any definite instructions," etc., and expressing confidence in his zeal and energy, and Terry's unwillingness to hamper him with precise directions, when nearly in contact with the enemy, did not warrant Custer in disobeying his orders. *It was only to govern his conduct when he should be in contact with the enemy*, in which case, of course, he would have to be the sole judge of what was best to be done.

* Journal of the Military Service Institution in the United States, Vol. XVIII., No. LXXIX.: "The Campaign Against the Sioux in 1876," by Major-General Robert P. Hughes, U. S. A.

His conduct in that case will be considered later. In any event it has no bearing on the question of disobedience, for the crux is here: had Custer obeyed orders, he would not have come in contact with the enemy when and where he did. The conditions would have differed greatly.

Every student of military matters knows that the words used, "He desires that you should conform to them (his own views) unless," etc., convey a direct, positive command.*

The abstract question of disobedience of orders is one that has often been discussed. It is impossible to maintain the position that an officer should never, under any circumstances, disobey his orders. Circumstances sometimes compel him to do so. But when an officer commanding troops which are supposed to act in coöperation with other troops receives orders to carry out a certain specified detail of a stated general plan, and in the exercise of his own discretion concludes to disobey his orders and do something other than what he was directed to do, he takes upon himself the onus of success or failure, not merely of his own immediate manœuver, but of the whole general plan. If the plan miscarries

*In Terry's report to the Secretary of War, under date of November 21, 1876, he gives his own understanding of his orders, which is fully warranted, in the following paragraph:

"At a conference which took place on the 21st between Colonel Gibbon, Lieutenant-Colonel Custer, and myself, I communicated to them the plan of operations which I had decided to adopt. It was that Colonel Gibbon's column should cross the Yellowstone near the mouth of the Little Big Horn, and thence up that stream, with the expectation that it would arrive at that point by the 26th; that Lieutenant-Colonel Custer with the whole of the Seventh Cavalry should proceed up the Rosebud until he should ascertain the direction in which the trail discovered by Major Reno led; that if it led to the Little Big Horn it should not be followed, but that Lieutenant-Colonel Custer should keep still further to the south before turning toward that river, in order to intercept the Indians should they attempt to pass around to his left, and in order, by a longer march, to give time for Colonel Gibbon's column to come up."

through his disobedience, whatever may have been his motives, woe be unto him! If by his disobedience he brings about the end at which the original plan aimed, the defeat of the enemy, that is another proposition. The event has then justified his disobedience.

Every soldier understands that reasons for disobedience must be so clear, so convincing, and so unexpected, that he is warranted in taking so prodigious a risk. Disregarding for the moment, for the sake of argument, General Ludlow's testimony as to preconceived and deliberate intent on Custer's part to disobey, supposing Custer's disobedience to have been caused by some exigency or crisis, we may ask ourselves what were the reasons that caused him entirely to disregard Terry's plan and so to manœuver as to bring himself directly in touch with the Indians in the shortest possible time, without attempting either to examine Tullock's Creek* or to incline to the southward—"feel with his left"? These reasons—if any there were—can never be known, owing to Custer's death. It can only be said that no satisfactory reasons appear which justify Custer's action.

The best that can be urged in defense of Custer is contained in the following paragraph taken from Colonel Godfrey's *Century* article.†

"Had Custer continued his march southward—that is, left the Indian trail—the Indians would have known of our movements on the 25th and a battle would have been fought very near the same field on which Crook had been attacked and forced back only a week before; the

* I have not discussed the Tullock's Creek matter. It is not material, except that the failure to examine it and to send a scout to Gibbon—some of Gibbon's men had been detailed with Custer for the purpose—with a report, was simply a further disobedience, and is, perhaps, a confirmation of Custer's deliberate purpose.

† *Century Magazine*, Vol. XLIII., No. 3: "Custer's Last Battle," by Colonel E. S. Godfrey, U. S. A.

Indians would never have remained in camp and allowed a concentration of the several columns to attack them. If they escaped without punishment or battle, Custer would undoubtedly have been blamed."

It may be pointed out with due reverence to Colonel Godfrey—whom I consider one of the ablest officers in the United States Army, by the way—that it is hard to see how Custer could have been blamed for obeying his orders, and that it is by no means certain that the Indians would have discovered Custer's column. Indeed, his previous success in concealing his movements and surprising the Indians (witness the Washita campaign) leads me to believe that he could have carried out his orders without observation. If Gibbon had struck the Indians first and had held them in play Custer could have annihilated them. General Fry's comments in the *Century* (appended to Colonel Godfrey's article) on Custer's action are entirely wrong.

As to what would have happened if Custer had been successful, it is more or less idle to speculate. Certainly, if he had overwhelmingly defeated the Indians, I do not think he would have been court-martialed; but if he had been in Reno's place and had been besieged with heavy loss, then I feel certain that Terry would have been in duty bound to prefer charges against him. All this is beside the main question, however, and it is now time to return to the history of the expedition.

Terry offered Custer four troops of the Second Cavalry and two Gatling guns, which were refused. Custer said that any force that was too big for the Seventh Cavalry alone to deal with would be too big for the Seventh Cavalry plus the four troops, and urged that the guns would hamper and harass his movements. Terry, who elected to go with Gibbon's infantry column, agreed with him.

Neither Terry nor Custer nor any one expected to meet more than one thousand warriors. They had no knowledge whatever of the large numbers of the so-called peaceable Indians, for whom rations had been regularly issued, who had broken away from the agencies and joined the hostiles. They did not know of Crook's defeat, and the great effect it had in inducing wavering bucks to give their allegiance to the brave men on the war-path. It will, perhaps, be fair to estimate the number of Indian warriors in the field at a mean between the white and Indian accounts, which range from twelve hundred on the one hand to three thousand on the other. To be on the safe side, I shall call it at least two thousand.* Whatever their number, there were enough of them.

In their way they were two thousand of the fiercest and most desperate fighters on the face of the globe. While they were undisciplined, untrained, and not entirely amenable to one will, as were the soldiers, they were, nevertheless, a fearfully formidable force. Their common hatred of the white man gave them sufficient coherence to form a rude but effective organization. They were led by experienced chiefs and were used to fighting. From 1868, after the close of the treaty by which the frontier posts were abandoned and the country restored to the Sioux and the Cheyennes, to 1876, no less than two hundred distinct fights, like that described in the account of the Yellowstone expedition, had occurred between the soldiers and the Indians. They were now to be tried in a real battle, and, as we shall see, they were not found wanting; for, in the end, all the honors of the campaign rested with them.

* Personally, I believe there were many more.

III. Custer's Expedition

The Seventh Cavalry left the camp at the mouth of the Powder River at twelve o'clock noon, on the 22d of June, 1876. Generals Terry, Gibbon, and Custer reviewed it as it marched away. With the column were fifty Arikara ("Rees") Indian auxiliaries, a few Crows, and a number of white scouts and newspaper correspondents. At four o'clock, after they had progressed twelve miles, the march was halted, and that evening the officers were summoned to Custer's headquarters, and marching instructions were given them. No bugle-calls were to be sounded. The march was to be made with the greatest possible rapidity; every officer was to look carefully to his men and horses. Squadron and battalion formations were abandoned; each troop commander was to report to Custer in person.

Custer was usually very uncommunicative. Ordinarily, he kept his plans to himself until the time to strike arrived. On this occasion, however, he announced his purpose, which was to follow the trail until they found the Indians, and then "go for them." He was not "carried away" by anything, and this declaration is further evidence of his deliberate purpose. His manner, at all times blunt and peremptory, not to say brusque, was now entirely changed. He was usually full of cheerfulness and confidence. There appeared to be a marked absence of both qualities in this instance. Officers have recorded that he seemed worried and depressed. It may be that he was feeling the displeasure of Grant, which his imprudent conduct had brought about. Perhaps the serious character of the risk he was taking by his independent move weighed upon him. If he succeeded, he would regain all he had lost in the censure. If he failed — well, he would not anticipate that. It was enough to give a man serious thoughts. His letters to his wife seem as cheerful and confident as ever, but, perhaps, he may have affected that for her sake. At any rate, the testimony as to his mental condition is unequivocal.

However he may have felt, he acted with his usual energy. Starting at five on the morning of the 23d, the regiment went into camp at five in the afternoon, having covered thirty-three miles over an execrable marching country — the "Bad Lands." On the 24th they marched twenty-eight miles over an even worse territory. Indian signs were abundant. Hundreds of Indians evidently had passed. As no one could tell how near they were to the hostiles, after supper on the 24th fires were put out and the men were allowed to sleep until half after eleven, while the officers and scouts examined the trail. It was reported to Custer that it led straight across the divide separating the Rosebud and the Little Big Horn. At half after eleven the men were routed out and marched ten miles toward the crest of the Little Big Horn Mountains, which they reached at two o'clock in the morning of Sunday, the 25th. A further halt was made, and at eight o'clock the advance was taken up once more.

They marched ten miles farther, and concealed themselves in a large ravine near the devide and about sixteen miles from the little Big Horn, about half after ten in the morning. Smoke was seen trembling in the air by the scouts in the crow's nest on the top of the divide, and there were other evidences of Indians down the valley of the Little Big Horn. It is believed that Custer intended to remain in hiding during the day, and deliver his attack on the next morning. Unfortunately,

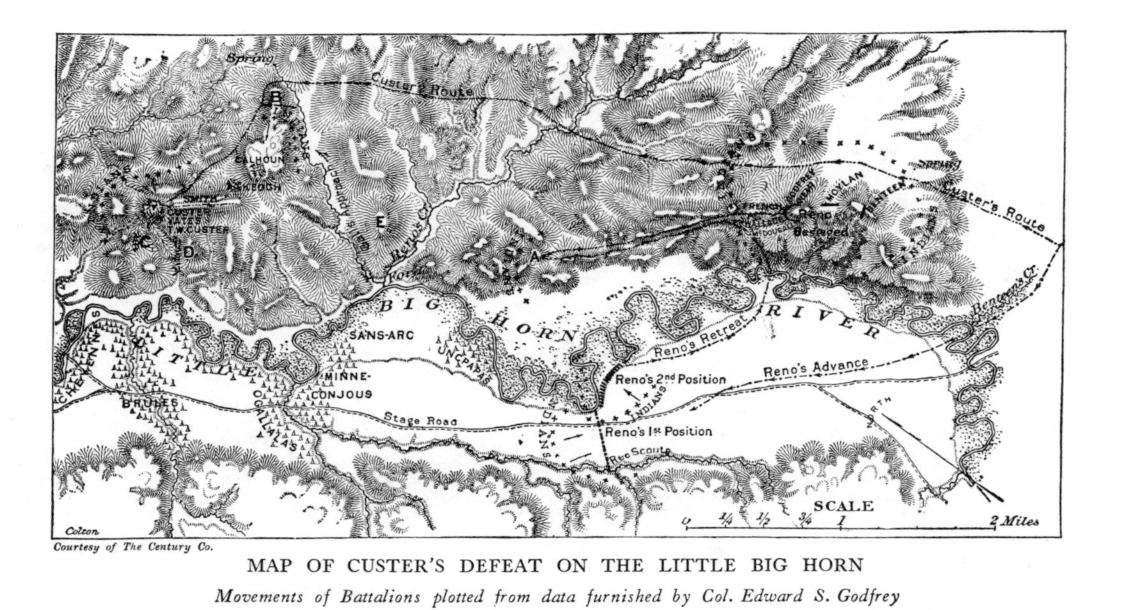

Courtesy of The Century Co.

MAP OF CUSTER'S DEFEAT ON THE LITTLE BIG HORN

Movements of Battalions plotted from data furnished by Col. Edward S. Godfrey

however, his trail had been crossed by the Indians. A box of hard bread had fallen from one of the pack-mules during the night march. When its loss was discovered, a squad of men had been sent back for it. They found an Indian trying to open it. He made his escape, and would undoubtedly alarm the villages they were approaching.

And now we come to another problem. As the result of his disobedience he was now practically in contact with the enemy, although he should not have been. Being in contact, however, what was he to do? There were no orders to govern him now. He was thrown on his own resources — just what he wanted, and what he had schemed and planned for. How was he to deal with his self-created opportunity?

Believing, as he and every one else did, that the Indian force did not greatly outnumber his own, an attack was entirely feasible. Should he deliver that attack, or should he wait to be attacked? The advantage is usually with the attacking party in Indian warfare. Should he seize or yield that? Suppose he decided not to attack the Indians, and they moved away and escaped? Would he not be censured for allowing them to get away, since he had got in touch with them?

Suppose — remote contingency — he were not entirely successful in his attack on the Indians? Gibbon must be somewhere in the vicinity. A day or two would probably bring him to the rescue. Could he not fight a waiting battle, if necessary, until the other column arrived on the field? Was it not absolutely incumbent upon him to embrace the opportunity presented to him? He had what he believed to be the finest regiment of cavalry in the service. He had tried it, tested it, on many fields; he knew, or thought he knew, the temper of his officers and men. He decided to attack. Indeed, there was nothing else for him to do. Fight he must. In the opinion of distinguished military critics who have expressed themselves upon the point, from General Sheridan down, he was justified in his decision. In that opinion I concur. And there is no evidence that he ever contemplated doing anything else. He had arranged matters to bring about the opportunity, and he had no hesitation in embracing it. Evidently, he had absolutely no premonition of defeat or disaster.

A little before noon he communicated his intention to his officers and men. He divided his regiment into three battalions. To Major Marcus A. Reno,* an officer with no experience in Indian fighting, he gave

* As the conduct of Major Reno was so decisive in the subsequent fighting, and since, upon his conduct as a pivot, the fortunes of the day turned, it is well to say something of his record, which I have compiled from official sources.

He was graduated from West Point in 1857, and was immediately appointed to the First Dragoons, and had risen to a captaincy in the First Cavalry at the outbreak of the Civil War. His career during the war was one of distinction. He was brevetted major, March 17, 1863, for gallant and meritorious services at Kelly's Ford, and lieutenant-colonel for gallant and meritorious services at the Battle of Cedar Creek, October 19, 1864. On January 1, 1865, he was appointed colonel of the Twelfth Pennsylvania Volunteer Cavalry, and was brevetted brigadier-general of Volunteers at the close of the war. Here is a brave and honorable record. Would that it might never have been tarnished!

He joined the Seventh Cavalry December 19, 1869, as major. He had had no Indian service prior to that time, and his services up to the present campaign comprised a three months' scouting expedition in Colorado in the summer of 1870. In 1879, upon his own application, a court of inquiry was convened for the purpose of investigating his conduct at the Battle of the Little Big Horn. It was the opinion of the court that no further proceedings were necessary in the case. One sentence of the record is significant: "The conduct of the officers throughout was excellent, and while subordinates in some instances did more for the safety of the command by brilliant displays of courage than did Major Reno, there was nothing in his conduct which requires animadversion from this court."

His relations with General Custer had not been friendly; so inimical were they, in fact, that Custer was begged, before starting on the fatal campaign, not to intrust the command of any supporting movement to Reno. Custer refused to allow any such personal considerations to prevent Reno receiving the command to which his rank entitled him.

In 1880 Major Reno was found guilty, by a general court-martial, of conduct unbecoming an officer and a gentleman. While in an intoxicated condition he had engaged

Troops A, G, and M; to Captain Benteen, a veteran and successful Indian fighter, Troops D, H, and K; Captain McDougall, with Troop B, was ordered to bring up the mule train and take it in charge; Custer himself took the five remaining troops, C, E, F, I, and L.

They left the ravine, and about noon crossed the divide which separated them from Little Big Horn Valley. Benteen was ordered to swing over to the left and search the country thoroughly in that direction, driving any hostiles he might come across into the village and preventing any escape of the Indians to the southward and westward. Reno was to follow a small creek, sometimes called Reno's Creek, to its junction with the Little Big Horn and strike the head of the village, supposed to be there. Custer's movements would be determined subsequently, although for the present he followed Reno. McDougall came last, following their trail with the slow-moving train, which dropped rapidly to the rear as the others proceeded at a smart pace. Benteen at once moved off to the westward, while Reno, followed by Custer, started down toward the valley of the Little Big Horn.

This river is a rapid mountain stream of clear, cold water, with a pebbly bottom, from twenty to forty yards wide. The depth of the water varies from two to five feet. While it is very tortuous, the general direction of the stream is northward to the Big Horn, which flows

in a brawl in a public billiard saloon, in which he assaulted another officer, destroyed property, and otherwise conducted himself disgracefully. The court sentenced him to be dismissed from the military service of the United States. The sentence was approved by President Hayes, and Major Reno ceased to be an officer of the Army in April, 1880.

It is painful to call attention to these facts, especially as Major Reno has since died; but the name and fame of a greater than he have been assailed for his misconduct, and in defense of Custer it is absolutely necessary that Reno's character and services should be thoroughly understood. For a further discussion of Major Reno's conduct, see Appendix B.

into the Yellowstone. The valley, from half a mile to a mile in width, is bordered by the bare bluffs. Along the river in places are thick clumps of trees. The Indian camp, the end of which they could see as they crossed the divide, was strung along the valley for several miles.

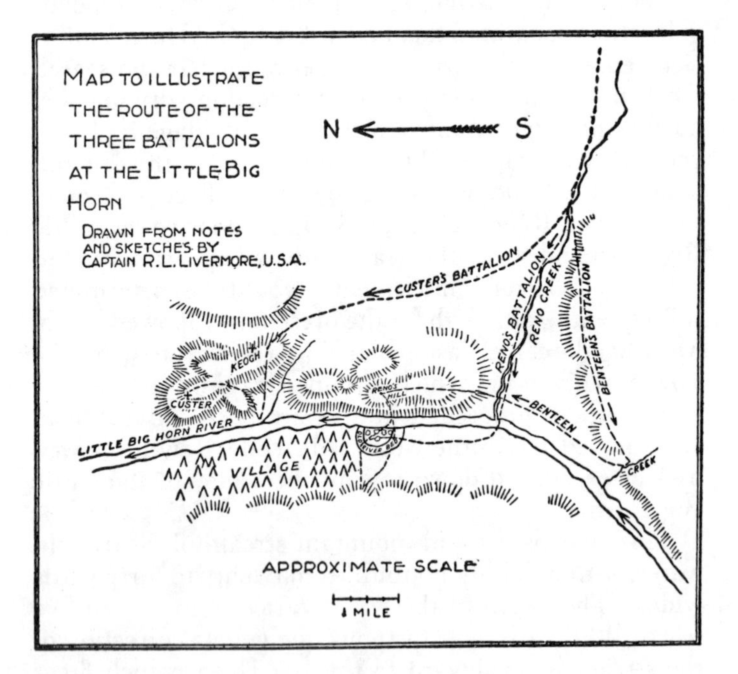

Reno's advance down the creek took him near to the east bank of river. Custer had followed him, slightly on his right flank. When Reno discovered the head of the village in the valley, he crossed the creek to Custer and reported what he had seen. Custer directed him to cross the river, move down the valley, and attack in force, informing him that he would be "supported" by

Custer's battalion. Reno accordingly put his battalion to a fast trot in columns of four, crossed the Little Big Horn River beyond the mouth of the creek, and proceeded onward for perhaps half a mile. Then he threw his troops in line, reaching from the river to the bluffs on the left, with the Arikara scouts on the left flank, and galloped down the valley for a mile farther.*

Reno stated subsequently that he believed that Custer intended to keep behind him all the time; and he fully expected, should he come in contact with Indians, that Custer would be on hand to join in the attack. Custer, however, had not continued down the creek or crossed the river with Reno, but had swung off to the high bluffs on the right bank of the creek, east of the river. Reno mistook the purport of Custer's statement. In order to support an attack, it is not necessary to get behind it. A flank attack or a demonstration in force, from some other direction, frequently may be the best method of supporting an attack. Custer's plan was entirely simple. Reno was to attack the end of the village. Benteen was to sweep around and fall on the left of it, Custer on the right. The tactics in the main were those which had been used so successfully in the Battle of the Washita (q.v.), and were much in vogue among our Indian fighters during the Indian wars.

Dividing forces in the face of an enemy to make several simultaneous attacks is dangerous, because it is almost impossible to secure a proper coöperation between the attacking units. A skilful general will concentrate his force upon the separately approaching and more or less isolated units and beat them in detail. Washington's tactics at Germantown were similar to

* Sure proof that the horses were not, as is sometimes urged, utterly worn out by the hard marching.

those of Custer; and his force, which would have swept the British from the field if his plans had been carried out, was beaten in detail for lack of coördination in the separate attacks. Some of Napoleon's most brilliant battles were fought when he occupied interior lines and by successive attacks broke up converging columns.

Still, the Indians were not believed to be veteran tacticians, although everybody underestimated their qualities. They were extremely liable to panic. A sudden attack or a surprise almost always disorganized them and threw them into confusion. Under the peculiar circumstances, I think there is little question that Custer's tactics were entirely sound and well considered, although this conclusion is often disputed. Where Custer made a mistake appears to be in his failure to take greater precautions that the attacks should be delivered simultaneously. He had a much longer distance to go than Reno and over a much worse country before he could attack, and he was not at all sure as to where Benteen was or when he could join. Nevertheless, the chances of success were many, the chances of failure few, and I have no doubt that Custer would have been successful had there not been a woeful lack of conduct on the part of his principal subordinate.

CHAPTER FIVE

The Last of Custer

I. Reno's Failure at the Little Big Horn

IT will be necessary, in order clearly to comprehend the complicated little battle, to treat each of the three operations separately, and then see how they were related to one another.

As Reno's men trotted down the valley, they saw, some distance ahead of them and to the right across the river on a line of high bluffs, Custer attended by his staff. The general waved his hat at them encouragingly, and disappeared over the brow of the hill. That glimpse of Custer, standing on that hill with outstretched arm gallantly waving his troopers on to battle, was the last any one of his comrades in the valley had of him in life; and it is certain that Reno must have realized then that Custer was not following him, and that he was expected to attack in his front alone.

However, Reno, having drawn near to the village, deployed his skirmishers, and slowly advanced down the valley. In a few moments they were hotly engaged with a constantly growing force of Indians.

Now, one thing about the battle that followed is the utter unreliability of the Indian reports of their movements. It is alleged that fear of punishment made them and keeps them reticent and uncommunicative. Different Indians tell different stories. Most of these stories disagree in their essential details, and it is impossible to reconcile them. It may be that the faculties of the Indians are not sufficiently alert to enable them to recall the general plan of the battle, or at least to relate it, although they knew well enough how to fight it at the time. Their accounts are haphazard to the last degree. Some say that they knew nothing of the advent of the troops until Reno's men deployed in the valley. At any rate, they had sufficient time, on account of his dilatory and hesitating advance, to assemble in heavy force. Reno had less than one hundred and fifty men with him. Even if Dr. Eastman's estimate,* that the Indians numbered but twelve hundred warriors, be true, they still outnumbered Reno, although, owing to the fact that the villages were strung along the river for several miles, only a portion of them were at first engaged with the troops. Flushed with their previous victory over Crook a short time before, these Indians now fell upon Reno like a storm.

Reno's line extended clear across the valley, which was quite narrow where the battle was joined, the right flank protected by the river, the left by the bluffs. Recovering from their alleged panic, possibly because of the feeble advance of the soldiers, the Indians rallied, and with wonderful generalship massed their attack on the left flank, which was most unfortunately held by the Arikara scouts. No Arikara that ever lived was a match for the Sioux or the Cheyennes. The Rees, as these Indian auxiliaries were called, broke and fled in-

* Charles A. Eastman, M. D., a full-blooded Sioux, a graduate of Dartmouth and the Boston University School of Medicine, who has published an interesting account of the battle from his investigations among the Sioux. See *The Chautauquan*, Vol. XXXI., No. 4, July, 1900.

CAPT. MYLES MOYLAN
LIEUT. A. E. SMITH *

MAJ. MARCUS A. RENO
CAPT. EDWARD S. GODFREY

SOME OF CUSTER'S OFFICERS

* Killed with Custer at the Little Big Horn

continently. They never stopped until they reached the supply camp on the Powder River, miles away. At the same time the horses of two troopers in the command ran away with them, and plunged straight into the Indian lines with their riders. Their fate was plain.

As the Ree scouts broke, the Indians turned Reno's left flank. The troopers gave way at once. There was no reserve which could be thrown upon the Indians until the line was restored. The whole force was slammed back, like a door, into the timber on the bank of the river.

Here Reno made a serious mistake. After rallying his men, he ordered them to dismount. Cavalry may be dismounted for defense, but sound judgment and military usage demand that for an attack, especially upon an Indian village of that kind, they should charge upon horseback. As one veteran cavalryman has written me, "I never could understand why Reno did not charge desperately on the Indians in front of him. His dismounting his men was against all sound military judgment. 'Audacity, always audacity,' is the motto for a cavalryman."* Had Reno been governed by this principle and charged, as he should have done, the result would have been different.†

The position was instantly surrounded by yelling Indians galloping madly to and fro, firing upon the troops. So far, Reno had lost but one wounded and the two who had galloped into the Indian line. His second position was admirable for defense. Sheltered by the trees, with his flanks and rear protected by

* General G. A. Forsyth.

† The unanimous testimony of the Indians who have discussed the battle subsequently is that they were panic-stricken by Reno's approach, and would have fled if his attack had been pressed home. This is about the only statement upon which the Indians all agree.

the river, he could have held the place indefinitely. However, he had not been detailed to defend or hold any position, but to make a swift, dashing attack. Yet, after a few moments of the feeblest kind of advance, he found himself thrown on the defensive. Such a result would break up the most promising plan. It certainly broke up Custer's. In spite of the defection of the Rees, a vigorous countercharge down the valley would have extricated Reno and might have saved Custer.

It is a painful thing to accuse an army officer of misconduct; yet I have taken the opinion of a number of army officers on the subject, and every one of them considers Reno culpable in a high degree. One at least has not hesitated to make known his opinion in the most public way. I am loath to believe that Major Reno was a coward, but he certainly lost his head; and when he lost his head, he lost Custer. His indecision was pitiful. Although he had suffered practically no loss and had no reason to be unduly alarmed, he was in a state of painful uncertainty as to what he should do next. The soldier, like the woman, who hesitates in an emergency which demands instant decision, is lost.

How long the troops stayed under the trees by the river bank cannot be determined accurately. Some have testified that it was a few moments, others an hour. Personally I think it was a few moments, which fear and apprehension lengthened to an impossible period. There had as yet been no panic, and under a different officer there would have been none; but it is on record that Reno at last gave an order for the men to mount and retreat to the bluffs. Before he could be obeyed, he countermanded this order. Then the order was repeated, but in such a way that nobody save those immediately around him heard it, because of the din of the battle then raging in a sort of aimless way all along the line, and no attempt was made to obey it. It was then repeated for the third time. Finally, as those farther away saw those nearest the flurried commander mounting and evidently preparing to leave, the orders were gradually communicated throughout the battalion, and nearly the whole mass got ready to leave. Eventually they broke out of the timber in a disorderly column of fours, striving to return to the ford which they had crossed when they had entered the valley.

Reno calls this a charge, and he led it. He was so excited that, after firing his pistols at the Indians who came valiantly after the fleeing soldiers, he threw them away.* The pressure of the Indians upon the right of the men inclined them to the left, away from the ford. In fact, they were swept into a confused mass and driven toward the river. All semblance of organization was lost in the mad rush for safety. The troops had degenerated into a mob.

The Indians pressed closely upon them, firing into the huddle almost without resistance. Evidently in their excitement the Indians fired high, or the troops would have been annihilated. The Indians supposed, of course, that they now had the troops corralled between them and the river, and that all they needed to do was to drive them into it. Chief Gall, who with Crazy Horse and Crow King was principally responsible for the Indian manœuvers, seeing the retreat of Reno to the river, summoned a large body of warriors, left the field and crossed the river farther down, intending to sweep down upon the other side and attack Reno's men as they struggled up the steep bank in case any of them succeeded in crossing. This was, as it turned out, a fortunate move for the Indians.

* This statement is elsewhere denied.

Meanwhile, Reno's men providentially found a pony trail which indicated a ford of the river. On the other side the trail led into a funnel-shaped amphitheater, surrounded by high, slippery bluffs. Into this *cul-de-sac* the whole fleeing body plunged, the Indians pressing the rear hard. The men jumped their horses from the bank into the water, and finding that the trail stopped at the bluff on the other side, actually urged them up the steep slopes of the hill.

There is no denying that they were panic-stricken. Although some of the veterans opened fire upon the savages, the bulk of the troopers did nothing but run. Dr. DeWolf was one of the coolest among those present. He stopped his horse deliberately, and fired at the Indians until he was shot dead. Lieutenant MacIntosh, striving to rally his men, was shot just as they left the timber. Lieutenant Hodgson, reaching the river bank, had his horse shot. In his agony the animal stumbled into the river and fell dead. The same bullet which killed the horse broke Hodgson's leg. He cried for help, and Sergeant Criswell rode over to where he lay. Hodgson took hold of the sergeant's stirrup, and under a heavy fire was dragged out on the bank, which he had scarcely reached before a second bullet struck him in the head, killing him instantly. Criswell was swept on by his men, but so soon as he could he rode back under a furious fire and brought off the body, as well as all the ammunition in the saddle-bags on several dead horses. He received a medal of honor for his courage.

If Gall had completed his projected movements, Reno's men would have been annihilated then and there. As it was, they reached the top of the bluffs without further molestation. They had lost three officers and twenty-nine men and scouts killed; seven men were badly wounded, and one officer, Lieutenant DeRudio, and fifteen men were missing.* These had been left behind in the confusion of Reno's "charge."

It was now somewhere between half after one and two o'clock in the afternoon, and during the fighting Reno was joined by Benteen's battalion. The Indians kept up a desultory fire on the position, but they seemed to have diminished in numbers. Reno occupied the next hour in reorganizing his force, getting the men into their accustomed troops, and taking account of casualties.

II. With Benteen's Battalion

In accordance with his orders, Benteen had moved off to the westward. He speedily became involved in almost impassable country, full of deep ravines, in which progress was slow and difficult. Water was very scarce in the country over which the regiment had marched until it reached the valley of the Big Horn. What water they had found that morning was so alkaline that the horses and mules, although they had been nearly a day without water, would not drink it. The horses were naturally tired, having marched over fifty miles since the morning of the day before, and the terrible up-and-down hill work exhausted them still more, although they were by no means played out. No Indians were seen by Benteen, and the condition of the country was such that it was evident there were none before him.

He turned to the right, therefore, and struck into the valley of the Big Horn, just ahead of McDougall and the pack train, intending to cross the river and attack the village or join Reno, as the case might be. He had

* DeRudio and one other man joined the command on the night of June 26th; the others succeeded in crossing the river to Reno's position late in the afternoon.

just watered his horses at a little brook following out a morass, when a sergeant from Custer's battalion passed by on a gallop, with a message for the supply train to come at once. As the trooper raced along the line he shouted exultantly, "We've got 'em, boys!" Benteen's men took this to mean that Custer had captured the village. A few moments after, Trumpeter Martini galloped up with a message from Custer to Benteen, signed by Cook, the adjutant, which read as follows:

"Benteen. Come on. Big village. Be quick. Bring packs.

"P. S. Bring packs."

The need for the spare ammunition with the pack-train was apparently so urgent that in his hurry Cook repeated the last two words. At the same time the sound of distant firing was heard in the valley. Making ready for instant action, Benteen led his troopers forward at a gallop down the valley. Tired though the animals were, they responded nobly to the demands of their riders, and the whole party swept across the hills in the direction whence the trumpeter had come until they overlooked the valley. Every one supposed that Custer had entered the valley and was driving the Indians before him. That he expected to have a big fight on his hands was indicated by the reiteration of his request that the pack train should be rushed forward, evidently to bring the reserve ammunition.

The valley was filled with dust and smoke; the day was frightfully hot and dry. Bodies of men could be distinguished galloping up and down. Benteen would, perhaps, have crossed the river and charged down the valley had his attention not been called to a body of men in blue on the bluff on the same side of the river to the right. They were, assuredly, hotly engaged, but there were also evidences of fierce fighting far down the valley. What was happening? What should he do? At this junction one of the Crow scouts — these Indians had not fled with the cowardly Rees, but remained with the command, fighting bravely — came up driving a small bunch of captured ponies, and he indicated that the principal battle was on the bluff. Benteen accordingly galloped around the bend of the river, and joined the demoralized Reno without opposition.

It is interesting to speculate what might have happened if Benteen had crossed the river and had charged down the valley. In that case, if Reno had recrossed the river and again attacked, the day might still have been won, but in all probability Reno would not have recrossed and Benteen would have been annihilated. At any rate, Benteen did the only thing possible when Reno's whereabouts and need were made known to him by the scout.

Reno had lost his hat in his famous "charge," and had his head tied up in a handkerchief. He was much excited, and apparently had no idea as to what he should do next. The officers of his battalion made no bones about admitting to the newcomers that they had been badly beaten and were in a critical condition. None of them could tell anything about Custer.

III. The Battle on the Bluffs

Benteen's men were ordered to divide their ammunition with Reno's. A line of skirmishers was thrown out around the bluffs, and an effort to get water from the river was made, the supply in the canteens having been long ago exhausted. The Indian fire prevented this. There was, of course, not a drop of water on the bluffs, and the wounded suffered greatly, to say nothing of the thirsty men. The officers collected in groups on the edge of the bluffs overlooking the field, and discussed the question. They were not molested by the Indians at this time.

The general impression was that Custer had made the mistake of his life in not taking the whole regiment in together. Possibly Reno's men took that view because they had been so badly mauled themselves. The valley had been filled with Indians, but, about three o'clock or a little after, most of them galloped down the river and were soon out of sight. The river banks were still lined with Indians under cover, who kept up a smart fire on Reno's men if they attempted to descend the bluffs and approach the water; but the main force had evidently withdrawn.

Firing was heard far away to the northward. It was heavy and continuous. There could be but one explanation of it. Custer's detachment had at last met the Indians and was engaged. This should surely have been a stimulus to Reno. Custer was fighting; Reno was not menaced—what should he do? Later in the afternoon two heavy volleys in rapid succession were remarked. This was so unusual under the circumstances that it was finally felt to be a signal from Custer. He must surely be in grave peril, then, and calling for help. How, in the name of all that was soldierly, could such an appeal be neglected? Many and anxious were the questions the officers and men put among themselves as to why Reno did not do something. It was felt by everybody that Custer was in grave jeopardy, and that Reno should move at once. He had about three hundred men under his command, one-half of whom had not been engaged.

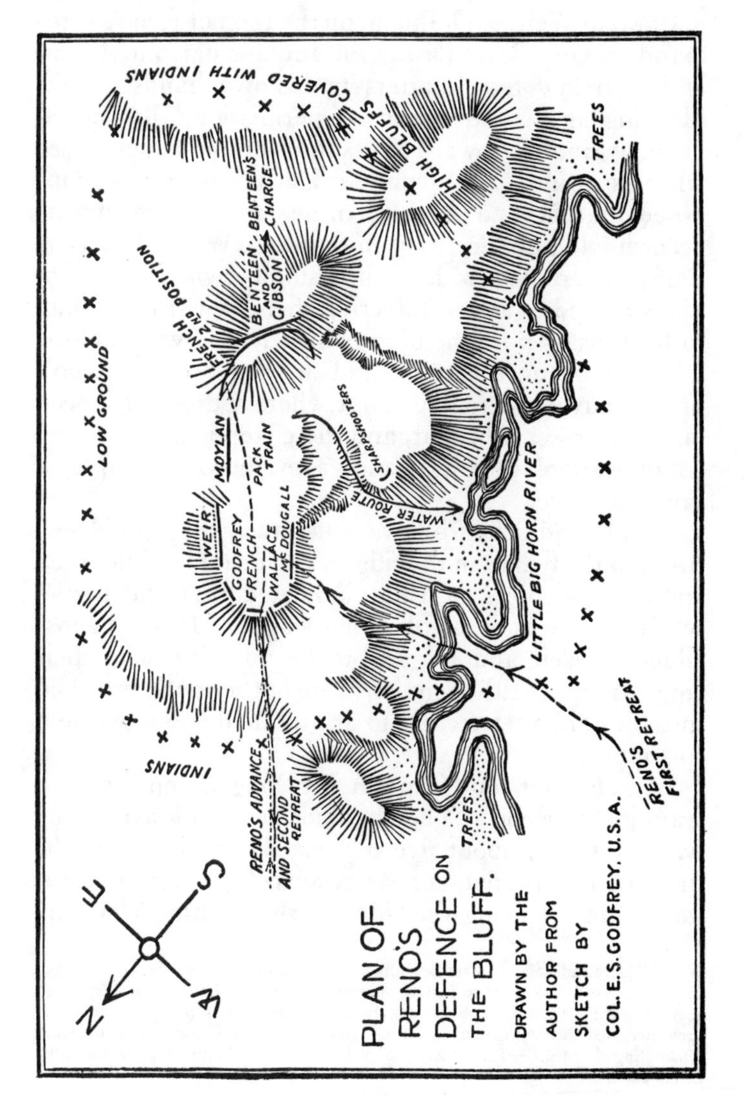
PLAN OF RENO'S DEFENCE ON THE BLUFF.
DRAWN BY THE AUTHOR FROM SKETCH BY COL. E. S. GODFREY, U.S.A.
N
E
S
W
INDIANS
LOW GROUND
COVERED WITH INDIANS
HIGH BLUFFS
TREES
BENTEEN'S CHARGE
BENTEEN AND GIBSON
FRENCH 2ND POSITION
MOYLAN
PACK TRAIN
SHARPSHOOTERS
WATER ROUTE
WEIR
GODFREY
FRENCH
WALLACE
McDOUGALL
LITTLE BIG HORN RIVER
RENO'S ADVANCE AND SECOND RETREAT
RENO'S FIRST RETREAT
TREES

Captain Weir, of D Troop, on the right of Reno's command, having cleared away the Indians in front of him, at last boldly took matters in his own hands. After pleading again and again for permission,* he started alone without it toward the sound of the firing to see what he could. Lieutenant Edgerly, his second, supposed that he had received orders to advance, and he accordingly put the troop in motion. Weir was on the bluff, Edgerly lower down in a small ravine. The Indians moved to attack Edgerly, when Weir signaled him to lead his men up the bluff, which he did without loss. The troop, unsupported and in defiance of Reno's orders, advanced to the point where Custer had been last seen to wave his hat, and there stopped. The men could overlook the ridges and valleys beyond them for a great distance.

A mile and a half or two miles away they could see, through the defiles in the ridges, great clouds of mounted Indians. Reports of rifles indicated that the battle, whatever it was, was still being waged. It was impossible for Weir and Edgerly to do anything with their single troop. Although they were not seriously attacked in their bold advance, Reno at first made no movement to support them.

At half after four Captain McDougall and the pack train joined Reno. They had not been molested in any way. At last, about five o'clock, Reno yielded to the urgent and repeated representations of the angry officers, and marched along the ridge to the position Weir and

* "The splendid officers of the Seventh, who had followed Custer so faithfully, begged Major Reno to let them try to join the general. They cried like women, they swore, they showed their contempt of that coward, but the discipline of their lives as soldiers prevented them disobeying until it was too late. You know Colonel Weir and Lieutenant Edgerly tried."—Private letter to me from the wife of an officer who was killed in the battle.

CAPT. THOS. W. CUSTER
CAPT. GEORGE W. YATES

LIEUT. JAMES CALHOUN
CAPT. MILES W. KEOGH

SOME OF CUSTER'S TROOP COMMANDERS

All killed with him at the Little Big Horn

Edgerly had reached. He came up to this point at half after five. The firing on the bluffs far ahead was practically over. The Indians could still be seen and some shooting was going on, but there did not appear to be a battle raging. They learned afterward that it was the Indians shooting into the bodies of the dead.

It was evident to every one that whatever might have been done earlier in the afternoon, there was no use in advancing now. Indeed, the Indians came sweeping back in great force in front of Reno, and at once attacked him. There was nothing for him to do but retreat to the most defensible position he could find, and endeavor to hold his ground. Custer and his men, if they still survived, must be left to face as best they could whatever fate had in store for them. Reno accordingly retreated to the place on the bluff whence he had just come. Lieutenant Godfrey, of K Troop, the rear guard, without orders deployed and dismounted his men, and, ably seconded by his junior, Lieutenant Luther R. Hare, by hard fighting kept off the Indians till the retreat was safely made by the rest, whom he and his troopers succeeded in joining. It was well that he did this, for his coolness and courage saved the command.

There was a little depression back of a ridge, which afforded some cover for the horses and pack train. During the retreat an incident occurred worthy of mention. One of the pack mules, loaded with precious ammunition, broke away and galloped toward the Indian line. Sergeant Hanley, of C Troop, sprang to his horse and raced after it. Officers and men called to him to come back, but knowing how priceless was the ammunition, he persisted in his course. He succeeded in heading off the animal, which turned and ran parallel to the Indian line, along which he galloped under a perfect shower of bullets, none of which, fortunately, touched him. He captured the mule, and brought it back with the ammunition intact. For this exploit he received a medal of honor.

The men took position around the ridge, across the depression and on a hill to the right, so as to protect the packs and the field hospital from all sides except on the river side, where the height of the bluffs and the distance prevented any Indian attack from that direction. Benteen's Troop H was placed on the right. They were on top of the break of the ridge and were without cover, the ridges being entirely bare of trees. Farther off, to the right, Benteen's position was commanded by higher ridges. At first the brunt of the fighting fell on the left, but the Indians soon surrounded the position and the engagement became general. The men threw themselves on the ground, and dug rifle-pits with their knives, tin pans — anything they could get. The fighting soon became severe, but gradually slackened as darkness approached, and stopped at about nine o'clock at night. The village in the valley was the scene of triumphant revel that night, and the shouting of the Indians could plainly be heard on the bluffs.

The early part of the night was full of wild confusion, but before long the soldiers recovered their equanimity and set to work strengthening their position. They were now completely surrounded; but most of them were under cover except Benteen's men, whose position, as has been stated, was overlooked by higher ridges within easy range. At two A.M., contrary to their usual habit, the Indians opened fire, but no attack was made. The next morning the battle began again in grim earnest.

The Indians pressed the party closer and closer. Benteen's exposed line suffered more than any other position. That experienced fighter saw that the Indians were massing in front of him, evidently intending to deliver a charge. If it fell upon his single troop it would not be possible to withstand it, and the whole force on the hill would be taken in reverse and annihilated. His men had nearly exhausted their ammunition, several had been killed, and there were a number of wounded to be attended to.

Ordering Lieutenant Gibson to hold the line at all hazards, Benteen ran to Reno, explained the situation, and begged for a reinforcement. After much urging he succeeded in getting Troop M, Captain French, sent over to the hill. Then he entreated Reno to allow the two troops to charge. Reno hesitated. Benteen urged him again and again, pointing out that if something were not done immediately, the position would be rushed and the command wiped out. At last he wrung a reluctant permission from Reno. He ran back to his position on the hill, and not a moment too soon formed his men up for the charge, putting himself at their head.

"All ready now, men!" he cried gallantly. "Now's your time! Give 'em hell! Hip! Hip! Here we go!"

The Indians had also given the word to charge, but Benteen was too quick for them. Leading his men with splendid bravery, revolver in hand, he rushed at the Indians. There was a brief hand-to-hand mêlée and the Indians broke and fled. Reno, seeing the effect of Benteen's gallant dash, actually led out a portion of his command on the other side of the hill and drove back the Indians in that direction. Benteen's magnificent courage had saved the day for the present.

The fire having slackened somewhat about eleven o'clock in the morning, volunteers were called to get water for the command, especially for the wounded. The Indians swept the banks of the river with their fire, and the attempt was hazardous to a degree. Nineteen men offered their services. Four of the best marksmen — Geiger, Windolph, Voit, and Mechling, of Troop H — were detailed to cover the others by taking an exposed position on the brink of the bluffs overlooking the river, as near as they could get to it. The other fifteen, one of whom has told me about the attempt, carrying canteens and camp-kettles, but without arms, crawled down through the bushes and ravines to the open space on the bank of the river, and then, covered by the rapid fire of the four men stationed above them, dashed for the stream. The Indians, who were execrable shots, opened a heavy fire upon them, but the men succeeded in filling the vessels they had brought, and though many of these vessels were hit and some of the men wounded, none of them was killed. A scanty supply of water it was, but it was a godsend. These nineteen also received medals of honor.

At three o'clock in the afternoon the firing, which had been maintained intermittently since noon, finally stopped, and later the men on the hill saw the Indians withdrawing from the valley. They set fire to the grass to screen their movements, but about seven o'clock in the evening they were distinctly seen moving out with all their possessions toward the mountains of the Big Horn. Eighteen troopers had been killed on the hill, and fifty-two wounded.

IV. The Last of Custer

Now let us turn to Custer.

Nobody knows exactly what he did. The testimony

of the field is not clear, and the statements of the Indians are contradictory. Dr. Eastman, an educated Sioux, has investigated the subject among many of his people, and arrives at one conclusion; Colonel Godfrey, one of the troop commanders who was with Benteen, and who has subsequently examined the field in company with Benteen and other officers, taking the testimony of Chief Gall, holds another. According to Eastman, whose account agrees with the popular understanding, Custer attempted to ford the river at a place now called Reno's Creek,* and fall on what he supposed to be the rear of the village, but which was really the middle of the upper half, and was driven back to the hills, where the final tragedy took place.

Godfrey, on the contrary, says that Custer, from the point where he was last seen by Reno's men, had a view of the village for several miles, although not for its whole length; that he must have been confident that he had it below him then, and that he made a wide detour in order to fall on the rear of the village. It was from this point that he sent the hurry-up message to Benteen. When at last, having gone far enough, as he thought, to take the village in the rear, or what he supposed was the rear, he turned toward the river, and was at once met by the Indians in great force.

It was probably about half after two in the afternoon. Reno had been forced back and driven across the river. Chief Gall, it will be recalled, had taken a large body of men across the river to intercept Reno on the other side. Before he could move down to the right for this purpose, Custer's men suddenly appeared on the hills.

* Not the Reno's Creek referred to above, down which Reno marched to the Little Big Horn. This Reno's Creek may be seen in the largest map between Reno's final position and the Custer Hill, where the general's battalion was overwhelmed.

Custer's manœuvering had been fine, and his appearance was a complete surprise, which at first greatly alarmed the Indians. Gall, however, did not lose his head. Rightly judging that Reno was temporarily eliminated from the game, he at once determined to attack Custer. He sent word of the situation to Crazy Horse, who was pressing Reno. Leaving just enough warriors to make a demonstration before the demoralized Reno, Crazy Horse galloped headlong down the valley, followed by his men and joined by others from the far end of the village, who had as yet taken no part in the fighting. They too crossed the river at the point where a deep ravine concealed their movements and enabled them to obtain a position on Custer's right flank. A similar ravine enabled Gall to menace the left flank. The Indians were in sufficient force completely to surround Custer. In the twinkling of an eye he found himself attacked in front and on both sides. Instead of advancing, he was forced to defend himself against an overwhelming attack. The troops were dismounted, horses moved to the rear, and Custer's men occupied the ridges.

Calhoun's troop was posted on the left, followed by those of Keogh, Smith, and Yates, with Tom Custer's on the extreme right. The last three troops happened to have the best defensive position upon the highest hill. With them was Custer. The Indians attacked at once. Riding at full gallop along the front of the line on their ponies, they poured a heavy fire from their long-range rifles upon the soldiers, to which the latter made a brave, steady, but not very effective reply with their inferior carbines. Keogh's and Calhoun's horses were stampeded at the first fire.

The force menacing them was so great that Custer

dared not leave his position on the hills. To retreat was hopeless, to advance impossible. They must stand on the defensive and pray that the advance of Reno's command up the valley, which they probably hoped that Benteen would reinforce, would compel the withdrawal of the Indians from their front. They fought on, therefore, coolly and resolutely, husbanding their ammunition and endeavoring to make every shot tell on their galloping, yelling foemen. They were, I imagine, by no means without ultimate hope of victory. The Indians in their accounts speak of the cool, deliberate courage of numbers of the officers and men, whom they singled out for their bravery.

Yet the troopers suffered great loss as the afternoon wore on. Their ammunition began to run low, and the contracting, whirling circle of Indians drove them closer and closer together. The remaining horses of the other three troops were at last stampeded, and with them went all of the reserve ammunition. The situation had evidently become so serious that Custer, in the vain hope that Reno would understand his peril at last, fired the two volleys which have been referred to. It appears at this time that he must have endeavored to send a message to Reno, for the body of a solitary soldier, Sergeant Butler, was found after the battle at a point half way between Custer's and Reno's commands. A little heap of cartridges lay near his body, evidencing that he had sold his life dearly. The Indians were acute enough — so they say, and probably with truth — to pick out the officers with Custer, and the mortality among them was fearful. It was evident to all on the hill, as the afternoon drew toward its close, that they were doomed. It was hardly possible that a counter-attack by Reno would save them now, and there were no evidences whatever that he was anywhere in the vicinity.

"Where, in God's name," they must have asked themselves in their despair, "can Reno be?"

One of the Crow scouts has said — although his account is generally disbelieved — that he went at last to Custer, as yet unharmed, and told him that he thought he could get him away, and that Custer, of course, refused to leave the field. The Crow altered his appearance by draping a blanket about him so as to look as much like a Sioux as possible, and in the confusion of the fight got away safely.* He was the only human survivor of the field.† What occurred after is a matter of conjecture, based upon the contradictory and inadequate testimony of the Indians themselves.

Gall and Crazy Horse now determined to end the affair. Massing their warriors in the ravine, they fell on both flanks at the same time that Crow King and Rain-in-the-Face led a direct charge against the front of the thinned and weakened line. They swept over the little band of men, probably now out of ammunition, in a red wave of destruction. There was a fierce hand-to-hand struggle with clubbed guns, war-clubs, and tomahawks, and all was over. Some twenty or thirty men, without their officers, who had probably all been killed where they stood, for their bodies were found grouped around that of Custer on the highest hill, endeavored to break through on the right. They were slaughtered to a man before they reached the river. A few scattered bodies, here and there in different parts of the field, indicated

* It is believed that this man, who was named Curley, secreted himself in a ravine, before the fighting began, and stole away at nightfall.

† Captain Keogh's horse "Comanche," badly wounded, was found on the field the second day after the battle. His life was saved, and for many years he was the particular pet of the regiment.

that separate men had made futile dashes for freedom. But the bulk of the command was found just where it had fought, with the *troopers in line, their officers in position!* They had been beaten and killed. Not an officer or man lived to tell the story, but they had not been disgraced.

There, the second day afterward, Terry, with Gibbon, having relieved Reno's men, found them on the hills which they had immortalized by their desperate valor. They had been stripped and most of them mutilated. Custer's body was shot in two places, in the side and in the temple. It was not scalped or mutilated. Colonel Dodge, an authority on Indian customs, declares that if Custer's body was neither scalped nor mutilated, he is convinced that the general committed suicide. None of the officers with whom I have communicated who inspected the body is willing to indorse this statement; on the contrary. Therefore, I am sure Colonel Dodge must be in error. The Indians give no particular information as to Custer's death. All that is known is that his body was there with those of his brave men.

With Custer in that fight perished many gallant souls. His brother, Captain Tom Custer, was the only man in the United States Army who held two medals for capturing two flags with his own hands in the Civil War. Rain-in-the-Face had accomplished his terrible revenge, for after the battle he had cut open the breast of the brave young soldier and had eaten his heart. Calhoun, of L Troop, was Custer's brother-in-law. With him was young Crittenden, a lieutenant of infantry, who had sought an assignment with Custer for this campaign. Smith was the captain of E, the Gray Horse Troop. At the storming of Fort Fisher, after two color-bearers had been killed, he had led his regiment to the attack, colors in hand. His shoulder had been smashed by a musket ball in that attack. He could never afterward put on his coat without assistance. With him was young Sturgis. Yates, a veteran of the Civil War, was captain of F, the Bandbox Troop; and with him was Riley, the youngest lieutenant there. Keogh, of I Troop, the oldest soldier of them all, and not the least brave, had been an officer of the Papal Zouaves in early life. He had a gallant record in the Civil War, too. With him was Porter, and with the others who had done their parts were Cook, the adjutant, and Lord, the doctor.

Others worthy of note fell on that fatal field: Mark Kellogg, a newspaper correspondent; Charlie Reynolds, the famous scout; Boston Custer, the General's brother, who was civilian forage-master of the regiment, and Autie Reed, the General's nephew — a mere boy, who wanted to see something of life in the West and who had welcomed with joy his opportunity to make the campaign. Well, he saw it, poor fellow! Indeed, the Custer family was almost wiped out on that fatal Sunday.

Premonitions of disaster, such as loving women may feel, were in the air that afternoon. Back at Fort Abraham Lincoln, the devoted wife tells how the women of the garrison assembled in her quarters in an agony of apprehension. There were words of prayer. Some one at the piano started "Nearer My God to Thee," and the women tried to sing it, but they could not finish it. It was not until the 5th of July that they received the news that at that very hour their loved ones were dying on the hill.

V. After the Battle

On the morning of the 27th of June Terry and Gibbon rescued Reno. The next day the surviving troops of the regiment, with some individuals from the other

Courtesy of The Century Co.

LIEUT. H. M. HARRINGTON *
LIEUT. J. E. PORTER *
LIEUT. W. VAN W. RILEY *

ADJ. W. W. COOK *
LIEUT. J. STURGIS *

LIEUT. J. J. CRITTENDEN †
LIEUT. DONALD McINTOSH ‡
LIEUT. BENJ. HODGSON ‡

OFFICERS OF THE SEVENTH CAVALRY

All killed at the Little Big Horn

* Killed with Custer. † 20th Infantry, attached to Custer's command. Killed. ‡ Killed with Reno

command, marched to the scene of Custer's defeat to identify and bury the dead. The bodies upon the dry grass had all been stripped and left, white and ghastly save for the red stains of wounds. The bodies of Doctor Lord, Lieutenants Porter, Harrington, and Sturgis, with those of a number of men, were not recovered. What became of them is not known to this day. They may have been captured alive and taken by the Indians to the village, and there tortured to death and their bodies disposed of. This, however, is unlikely. The Indians positively deny that they took any prisoners, and it is probable that they did not. There are quicksands near the bed of the Little Big Horn, and possibly those bodies were engulfed in them. But all this is only surmise. No one can tell anything about it, except that they were undeniably killed. And we may be certain they died as brave men should.

They buried two hundred and twelve bodies on the hill, and the total losses of the regiment in the two days of fighting were two hundred and sixty-five killed and fifty-two wounded — over fifty per cent. The losses of the Indians were never ascertained. They did not, however, begin to equal those of the soldiers. It is grossly unfair to speak of the battle as the "Custer Massacre," as is often done. Custer attacked the Indians, and they fought him until all the white men were killed. There was no massacre about it.

The cause of the disaster must, first of all, be laid to Custer's disobedience of orders. In spite of that, however, I think it is probable that he might have won the battle, or at least made good his defense until relieved by Terry and Gibbon, although sustaining heavy loss, had it not been for three happenings. The first was the vastly greater number of Indians in the field than any one expected to encounter. The next, and to me this is absolutely decisive, was Reno's failure to press his attack. If he had gone in with the dashing gallantry which was expected of him, while it is certain that he could not alone have whipped the Indians, yet he could have so disorganized them as to have maintained his position in the valley in the midst of the village without the greatest difficulty, until Custer could fall upon the rear of those attacking him, and Benteen, with the pack train, could reinforce them both. The Indians say that they were demoralized for the time being by Reno's sudden appearance, and that the squaws were packing up getting ready for flight when the weakness of Reno's advance encouraged them to try to overwhelm him. Custer had a right to expect that Reno would do his duty as a soldier and take a bold course — which was, as usual, the only safe course.

Colonel Godfrey, in his account, suggests still a third cause. The carbines of the troopers did not work well. When they became clogged and dirty from rapid firing, the ejectors would not throw out the shells, and the men frequently had to stop and pick out the shells with a knife. The chambers of the carbines at that time were cylindrical, and the easily accumulated dirt on the cartridges clogged them so that the ejectors would not work properly. The chambers were afterward made conical, with good results. The Indians had no such trouble. Their weapons were newer and better than those of the soldiers. If the indifferent weapons of the troopers failed them, their annihilation in any event would have been certain.*

* It is possible that if Custer had kept the regiment together, he would have won the battle; but this is by no means certain, and authorities differ. I think he would have been forced eventually on the defensive.

I have censured Custer somewhat severely in this article, and it is a pleasure to me to close it with a quotation from Captain Whittaker's life of his old commander. In this quotation Lawrence Barrett, the eminent actor, who was an old and intimate friend of Custer, has summarized the character of the brave captain in exquisitely apposite language; and, in his words, I say good-by to the gallant soldier whose errors were atoned for by an heroic death in the high places of the field:

"His career may be thus briefly given: He was born in obscurity; he rose to eminence; denied social advantages in his youth, his untiring industry supplied them; the obstacles to his advancement became the stepping-stones to his fortunes; free to choose for good or evil, he chose rightly; truth was his striking characteristic . . . his acts found his severest critic in his own breast; he was a good son, a good brother, a good and affectionate husband, a Christian soldier, a steadfast friend. Entering the army a cadet in early youth, he became a general while still on the threshold of manhood; with ability undenied, with valor proved on many a hard-fought field, he acquired the affection of the nation; and he died in action at the age of thirty-seven, died as he would have wished to die, no lingering disease preying upon that iron frame. At the head of his command, the messenger of death awaited him; from the field of battle where he had so often 'directed the storm,' his gallant spirit took its flight. Cut off from aid, abandoned in the midst of incredible odds . . . the noble Custer fell, bequeathing to the nation his sword; to his comrades an example; to his friends a memory, and to his beloved a Hero's name."

NOTE.—The question concerning Custer's conduct is

so important a one that I have included in Appendix A the opinions, pro and con, of several officers with whom I have corresponded; and in which I have indicated some other sources of information by which the reader may settle the debatable question for himself.

CHAPTER SIX

One of the Last Men to See Custer Alive

MR. THEODORE W. GOLDIN, of Janesville, Wisconsin, formerly a trooper of the Seventh Cavalry, now Chairman of the Republican State Central Committee of Wisconsin, was the last, or perhaps the next to the last, man to see Custer alive. He has prepared an account of his personal experiences in the battle, which is one of the most interesting of the contributions that have been made to this volume. His description of the death of Hodgson is splendidly dramatic, as is his story of the brief conversations between Custer and Keogh, Reno and Weir, and Benteen and Wallace.

What a magnificent picture is that presented by Hodgson, determined to retreat no longer, facing about, drawing his revolver, and dying at last, face to the foe, weapon in hand! Mr. Goldin contradicts the popular impression, as repeated by Colonel Godfrey and others, that Reno threw away his pistols in his "charge."

Milwaukee, Wis., August 11, 1904.

My Dear Sir:—

I am in receipt of your letters of July 28th and August 2d, asking me for a few reminiscences of personal experiences and touching on my acquaintance and knowledge of Gen. Custer and his last fight.

The years that have elapsed since that stirring event may have somewhat dimmed my recollections, and the time at my disposal at this time is so very limited, that what I may say must, of necessity, be somewhat fragmentary. Your articles on this subject, as well as the entire series, have been read with great interest, and I am very much pleased to know they are to be published in book form.

In reply to your request, I will say that I had known Gen. Custer from the time I joined the regiment in 1873 up to the time of his tragic death, and had campaigned with him and with the regiment with the exception of the year 1875, when the troop to which I was attached was stationed in the South.

Early in the spring of '76, we received word that an expedition was being organized against the Sioux, and that three large columns were to take the field, and a few weeks later our marching orders came, and our battalion changed station from near Shreveport, La., to Fort Lincoln, D. T.

At the time of our arrival Gen. Custer was in Washington before some investigating committee, and only joined the regiment a few days before our column took the field, he having been ordered under arrest at Chicago while en-route from Washington.

I will not take the time or space to touch on our march from Fort Lincoln to the Yellowstone, which we struck near the mouth of the Powder River. Here we remained a few days while outfitting our pack train, as it had been determined that we would abandon our wagons here and establish a supply camp. Maj. Reno had left us a few days before on a scout, expecting to rejoin us near the mouth of the Tongue River on June 17th.

After arranging to leave our wagon train and some dismounted recruits and the regimental band at the Powder River, we moved camp to the mouth of the Tongue, where we lay on the 17th of June, the day on which Gen. Crook had his big fight with the Indians on the Rosebud. During the day we watched every distant dust cloud that whirled across the river bottom, hoping that it meant the approach of Maj. Reno's command, but as nothing was heard from him, we broke camp on the morning of the 18th and moved up the valley, where we effected a junction with Reno soon after noon. During the previous days we could not help but note the fact that Gen. Custer seemed moody and discontented, and, entirely different from his usual habit, appeared nervous and excited to some extent.

When Reno came in with his report he became a changed man. His old-time energy and snap were made apparent by the manner in which he hustled the command into marching order and took the trail for the Rosebud, where Gen. Terry on our supply steamer the "Far West" had already preceded us in the hope of striking the command of Gen. Gibbon, who was known to be somewhere in that vicinity. Our march was prolonged far into the night, but we finally struck the Yellowstone and went into camp, assured that if the steamer was below us we were bound to intercept it.

The following morning we were in the saddle early, and soon after noon we sighted the wagon train of Gen. Gibbon's command moving slowly up the river and a few moments later the "Far West" steamed into sight. Couriers were sent out to advise them of the fact that Reno had discovered the trail of a considerable body of Indians, and in a short time we were in camp near the mouth of the Rosebud and a council of war was held at which we understood that Reno's report was discussed in detail.

The following day arrangements were made for a vigorous campaign. A final council was held on the steamer on the night of the 21st, as I now recollect it, at which time Gen. Custer received his final orders, substantially as you have stated them. At this council, and just about the time it was breaking up, the question was asked Gen. Gibbon as to what time he could reach the mouth of the Little Big Horn River, and he replied,

"Not before noon of the 26th."

At this conference it was reported that Custer had been offered the battalion of the Second Cavalry and the artillery which had been brought up on the steamer, but declined both, claiming that he knew his command so well that he preferred to trust himself with them alone, and that he feared the artillery might delay his march when nearly in presence of the enemy. Many of the officers and men felt that it was but a part of a preconceived scheme to secure an independent command, such as he had been used to having for years, but be that as it may his request was granted.

On the afternoon of June 22d we passed in review before Gen. Terry and Gen. Gibbon, and soon struck the trail described by Maj. Reno. From that time until Reno struck the Indian village we did not deviate from it except when it became necessary to find a satisfactory camping ground. No attempt was made to scout the country as we had been directed to do, nor was any attempt made to send a courier across to Gen. Gibbon, although a man named Herndon, a scout, had been attached to our command for that purpose.

Our marches were long and our movements very rapid until the night of the 24th, when we moved off the trail some distance and apparently settled down for the night. But this was only a ruse to mislead the Indians, as we had received orders not to unpack our saddles and to be prepared for a night march.

From the hour we left the Rosebud Gen. Custer acted in many respects like another man, his old-time restless energy had returned, and he seemed to think of nothing but to reach and strike the Indians. In this connection it might be well to say that the trail we were following led from the direction of the Missouri River and indicated, according to the estimate of Bloody Knife and some of our Crow scouts, a band of from a thousand to twelve hundred Indians, whom we afterward learned came from the Missouri River agencies and consisted almost entirely of warriors.

About eleven o'clock we received word to saddle up and lead into line. In the meantime a scouting party of Crows and Rees, with some of our officers, Lieut. Varnum and, I think, Lieut. Hare, had pushed on ahead of us to scout the trail across the divide and seek to locate the Indians if possible. Leaving our bivouac, we again struck the trail and pushed forward, seeking to cross the divide and get into the shelter of the foothills along the Little Big Horn before daybreak the following morning. Owing to the roughness of the country and the difficulty in scouting the trail, we were unable to do this, and daybreak found us in a ravine at the foot of a range of high bluffs, just how far from the river we did not know.

Some time during the night it was said Gen. Custer

had pushed ahead and joined the scouts, and that just after daybreak they told him they had located the village in the valley of the Little Big Horn, but Gen. Custer replied that he did not believe them. In the meantime it became apparent that our presence in the country was known, as, during the night a box of ammunition* had been lost and a detachment sent back to recover it came across some Indians trying to open the box. On this being reported to Gen. Custer, he seemed to decide on an immediate advance.

In the talk, just about the time the division was made in the command, Mitch Bowyer, a half-breed Crow interpreter, said to Gen. Custer that he would find more Indians in that valley than he could handle with his command. Custer replied that if he (Bowyer) was afraid to go he could stay behind. Bowyer replied that he was not afraid to go wherever Custer did, or something to that effect, but that if they went in there neither of them would come out alive.

Just before the advance was made I was detailed by Lieut. McIntosh to report to Gen. Custer for duty as orderly, and at once did so. The general directed me to ride with Lieut. Cook, our regimental adjutant, and perform any duties he might assign to me.

During our brief halt the men had thrown themselves on the ground and were most of them asleep, while the horses were grazing among the sage brush. Gen. Custer ordered the advance, saying that the company in each battalion first ready should have the right of the line, and in a few minutes we were all in the saddle. During the halt it had been ordered that Benteen with

* Mr. Goldin's recollection is in error, according to Colonel Godfrey, who is positive that the box contained hard bread. However, the difference is not material —it was a box, anyway! — C. T. B.

SITTING BULL

Chief Medicine Man of the Sioux Nation

his battalion should move off to the left, scouting the country in that direction, driving before him any Indians he might discover, and sending word to the command of anything he might find. Reno was to follow the trail, while Gen. Custer with the five troops under him struck off to the right, leaving McDougall with the pack train to follow as best he could.

We immediately took up the line of march in accordance with these orders, and after probably a half hour's hard ride the impassibility of some of the hills and ravines forced the column under Gen. Custer to veer off to the left and we soon came up with the command of Maj. Reno, which was pushing ahead on the trail as fast as the roughness of the ascent would permit. After a hard climb we reached the top of the ridge, where we saw before us a rolling plateau sloping off toward the foot-hills of the river, which was perhaps some five or six miles away.

As our command dashed over the divide we could see Reno some distance in front of us, moving rapidly down the trail, while several miles to our left was the command of Col. Benteen, scouting the bluffs as he had been ordered. For some distance we followed the general course of Reno's advance, but were some distance in his rear and to his right.

To those of us who were near him it seemed that Custer was chafing at the apparent slowness of our advance, as he would at times dash ahead of the column and then rein in and await our approach and again off he would go. Just about this time we discovered a huge dust cloud moving down the river valley, but could not determine the nature of it. As soon as this was discovered Custer rode over toward the river accompanied only by his orderly trumpeter, and stopped for a moment on the top of a high pinnacle, where we saw him wave his hat, apparently in salutation to some one in the distance, and then come dashing back toward the head of the column which was headed by Capt. Keogh and "I" Troop, veterans of a dozen fights. With his eyes snapping in his excitement, Custer rode up to Keogh and said, somewhat excitedly:

"Keogh, those Indians are running. If we can keep them at it we can afford to sacrifice half the horses in the command."

Calmly as though on dress parade, Keogh turned in his saddle and looked back at the long line of eager, bronzed, bearded faces, and turned to Custer with the remark:

"General, we will do all that man and horse can do."

A moment later the bugles blared out the charge, the first bugle note we had heard since leaving the Rosebud, and away we thundered northward down the river, two or three times seeking to find a place where we could work down into the valley below us, but without success.

After perhaps the third unsuccessful trial, Gen. Custer talked hurriedly with Capt. Cook for a moment, and Cook pulled out his pad and dashed off a line or two, which he folded up, at the same time calling for an orderly. I happened to be the first one to reach him, and he handed me the paper with the order.

"Deliver that to Maj. Reno, remain with him until we effect a junction, then report to me at once," and he was gone.

An instant later the rear of the column dashed past me and was lost to sight in the ravines. For an instant I looked after them, and then realizing that I was in a dangerous country and alone, I lost no time in heading in the direction of Reno's command, which I was able to locate by the dust cloud that hovered over them. Fortunately, I was not molested to any great extent. A few long-range shots were fired at me, which only served to accelerate my speed and materially added to my desire to be among friends.

A ride of some five or six miles and I overtook Reno just as he was dismounting to fight on foot. I delivered my dispatch, the contents of which I did not know. Reno glanced at it somewhat hurriedly and stuck it in his pocket. About this time the Ree scouts stampeded and, as we afterward learned, did not recover their sand until they reached the Powder River and the shelter of the wagon train. It soon became apparent that the Indians were passing our flank and coming in behind us, and we were forced to face about and endeavor to repel their advances until we could get our horses into the timber, in which attempt several horses were shot and two or three stampeded. Soon after this we retired into the timber, where we had better protection and resumed the fight. Sheltered by the timber and the river bank, we were able to make a much better defense for a short time.

Soon after this we noticed that Capt. Moylan was mounting his troop and Lieut. Wallace, who stood near me on the skirmish line, called to Capt. French, who was commanding the center company, and asked what the orders were. French replied that he hadn't received any but would try and find out, and in a few minutes he called to us saying he understood they were going to charge, that he had not received any orders, but we might as well mount and support them.

We were ordered to get to our horses, and while doing this we found that some fifteen of our men were dismounted either because of the shooting or stampeding of their horses. Lieut. McIntosh had lost his horse and took one belonging to a trooper named McCormick, who gave him up with the remark that we were all dead anyway, and he might as well die dismounted as mounted. Swinging into the saddle, we moved out of the timber and to our surprise discovered that instead of "charging the Indians" Reno was executing a masterly charge on the bluffs on the opposite side of the river.

As soon as the Indians discovered this, they massed on our flanks and opened a heavy fire on the retreating column. Fortunately, they were poor marksmen mounted, and our loss was comparatively small at this stage of the stampede, for that is what it was.

It is reported that Reno became so excited that he emptied his revolver at the Indians and then threw the weapon from him. I happen to know this was not so, as the revolver is now in the possession of Gen. Benteen or his family, or was a few years ago.

During the progress of this retreat I was riding on the left of our column and near the timber, and when almost in sight of the river my horse fell, throwing me into a bunch of sage brush, but without doing me serious injury save to exterior cuticle. As I scrambled to my feet Lieut. Wallace passed me, shouting for me to run for the timber as my horse was killed. I did not stop to verify his report, but took his advice, striking only one or two high places between where the horse fell and the timber, which I presently reached.

From where I was concealed I could see our men force their horses into the river and urge them across the boulder-strewn stream. I saw Lieut. Hodgson's horse leap into the stream and saw him struggling as though wounded, I saw the lieutenant disengage himself from the stirrups and grab the stirrup strap of a passing trooper and with that aid make his way across the stream. No sooner had he reached the bank than it became apparent he had been wounded, but he pluckily held on, and the trooper seemed to be trying to help him up behind him on the saddle, but without daring to stop his horse. An instant later Hodgson seemed to be hit again, for he lost his hold, fell to the ground, staggered to his feet and sought to reach another comrade who reined in to aid him, and just as it seemed that he was saved I saw the second trooper throw up his arms, reel in the saddle and fall heavily to the ground. Hodgson started to make his way toward the ravine up which the command was disappearing, he staggered forward a few steps, stumbled, struggled to his feet again, only to fall once more. He apparently decided that further effort to retreat was useless, as I saw him turn and face the Indians, draw his revolver and open fire. An instant later three or four shots rang out from my side of the river, and I saw Hodgson reel and fall and I knew it was all over.

In the meantime our men had succeeded in crossing the river and made their way up a neighboring ravine, all save those who had met their fate at the ford, which was one of the worst along the river for many rods. Left alone, I began to wonder what my own fate was likely to be, but I was not observed and therefore not molested, the Indians being busy stripping and mutilating the bodies of our dead along the banks of the stream.

About this time I could hear sounds of heavy firing down the river, and made up my mind that Custer was engaging the Indians, and from the momentary glimpses I had of the village I felt that he was as badly outnumbered as we were. Most of the Indians in our front melted away and I could see them lashing their ponies as they hurried to join their friends at the lower end of the village.

About this time I saw the scout, Herndon, some little distance from me, making his way toward the river, and called to him, and we were soon together. He told me that the fifteen dismounted men of our outfit had made their way to a point in the timber about a couple of hundred yards from where we were, but that in order to get nearer the river they had to cross an open space and every time they tried it the Indians fired on them. Before leaving us the Indians had set the river bottom on fire, evidently with a design of concealing their movements, or of smoking or roasting out our wounded. This smoke proved to be our salvation, as under its cover we made our way to the river, forded it with some difficulty, and stumbled on to one of our Crow scouts, who pointed out to us the location of the command, which we soon joined.

We found that Benteen and his battalion had reached it and that nothing was known of the location of Custer and his command. We reported what we had heard and seen, and just about this time some one discovered a white flag waving from a point in the river bottom near where Herndon and myself had been concealed. Lieut. Hare, at the risk of his life, crept down to a point of bluffs overlooking the valley, and after considerable signaling satisfied the party we were friends and they made their way across the river and soon joined us, proving to be the dismounted men who had been left to their fate when Reno made his retreat.

I omitted to state that when we started on our retreat Lieut. McIntosh, mounted on McCormick's horse, was several rods in front of me and I noticed that in some way his lariat had become loosened and was dragging on the ground, the picket pin striking sage brush and other obstacles, and rendering his immediate vicinity very dangerous. Several of the men sought to call his attention to it, but evidently he did not hear them. A moment after this I saw his horse go down, but whether he was shot or not I do not know. All I could see as I passed was that the lieutenant was lying where he fell, and was either dead or stunned, probably dead, as we found him in the same place the day after the Indians left us.

As we were standing on the bluffs looking down into the valley I heard some loud talk near me, and turning in that direction, I heard Capt. Weir say:

"Well, by G—d, if you won't go, I will, and if we ever live to get out of here some one will suffer for this."

He strode away, and a few moments later I saw "D" Troop mount up and move down the valley in the direction of the distant firing. Apparently without orders, the entire command followed them in no sort of military order with the exception of the two troops under Benteen and Godfrey. In this way we pushed down the valley some distance, when we discovered Weir and his troop falling back before a largely superior body of Indians.

Hastily forming a line, we held back the advancing horde until Weir and his command had passed our lines and formed some distance in our rear, where, with the support of some of our men who were near them, they formed a line and opened fire, permitting us to fall back and re-form again in their rear. In this way we fell back some little distance, when Col. Benteen, who seemed to be the leader in our section of the field, spoke to Lieut. Wallace, saying:

"Wallace, there is no use falling back any further. Form your troop, your right resting here, and we will make a stand."

Wallace grinned and said,

"I haven't any troop, only two men."

Benteen laughed grimly and answered,

"Form yourself and your two there, and I will tell you more about it when I find out myself."

That was the nucleus of our line of defense. Others soon joined us and we sheltered ourselves behind sage brush and hurriedly heaped piles of dirt and opened fire, keeping the Indians at a respectable distance until darkness came to our relief. During the night we changed our position a trifle, located our corral and hospital, and put in the night intrenching ourselves as best we could. At daybreak the fight opened again and continued without intermission until about three in the afternoon, when to our surprise the Indians began to take down their tepees, pack their travois, and in a few hours were moving up the river valley, a great mass of ponies, travois and Indians, unfortunately just out of rifle range.

Twice during the afternoon volunteer parties had gone for water, each time being fired upon by the Indians, but it was only on the second trip that any one was hit. Poor Madden, of "K" Troop, was the unfortunate, his leg being shattered three times between the ankle and the knee. We carried him back to the hospital, where his leg was amputated that night.

On the following day Gen. Terry and Gen. Gibbon came to our relief, and through them we received the first authentic information as to the fate of our comrades of the other battalion. On the 28th, after having transported our wounded across the river, we visited the scene of the battle and buried such of Custer's men as we found. Aside from General Custer, we found hardly a body on the field that had not been mutilated in some manner or another, but as I recollect now, we found no marks of mutilation on our dead leader.

In the space at my disposal it is not possible to deal in incidents of the fight or go into detailed descriptions. In fact, those have been well covered in your article already.

Since the fight I have discussed it with many officers of the army, and others who have had experience on the frontier, and the general opinion seemed to be that there were two, possibly three, main causes for this disaster.

First: A division of the command into practically four separate columns while still some fifteen miles from the battle-field, and without accurate knowledge as to the exact location or approximate strength of the enemy, and the separation of those columns so that at a critical period of the fight no two of them were in supporting distance of one another.

Second: The fact that Custer came into the presence of the enemy practically twenty-four hours ahead of time.

Third: The loss of the horses and with them much of the surplus ammunition of Custer's command.

This subject has been so often discussed by men much abler than myself that I will not attempt it here.

My experiences with Gen. Custer always led me to look upon him as somewhat recklessly brave, disposed to take chances without fully considering the odds against him.

I have always felt that one possible reason for the course he followed, in the face of the orders he received, might be attributed to the fact that he was feeling keenly the apparent disgrace of the treatment accorded him by the President, and that he thought that by a brilliant dash and a decided victory, similar to his Washita fight, he might redeem himself and once more stand before the people as a leader and an Indian fighter second to none. Whatever may have been his motive, we must all admit that he made a most gallant fight and gave his life at the side of the comrades who had ridden with him to victory in many a previous battle.

Very truly yours,

THEO. W. GOLDIN.

CHAPTER SEVEN

The Personal Story of Rain-in-the-Face

BY

W. Kent Thomas

NOTE. — It is rare, indeed, to get the Indian side of a story in so clear, so connected, and so dramatic a form as is the following account of the Battle of the Little Big Horn from one who played a great part in it and in the events that led up to it. This is a unique document in our records, and is inserted here by kind permission of Mr. Thomas. It originally appeared in *Outdoor Life*, Vol. XI., No. 3, for March, 1903. Its accuracy and fidelity to fact are so attested as to be beyond question. — C. T. B.

THE writer saw much of the "Custer Indians" at the World's Fair and afterward at Coney Island, and had a good chance to know some of them well. The following leaves from a diary kept at that time show how the Indians regarded the Custer fight; they considered that the white men were simply outgeneraled by Sitting Bull:

Coney Island, N. Y., Aug. 12, 1894.

Rain-in-the-Face (Itiomagaju) hobbled into the tent to-night, as McFadden and I were discussing the events of the day, and seating himself, unbidden, with true Indian stoicism, he grunted out that one word of all words so dear to a Lakota, "Minnewaukan!" which, literally translated, means "Water of God," but which by usage has been interpreted as "fire-water." Since the other Indians were all away from camp on a visit to their friends, the Oglalas at Buffalo Bill's camp, I decided to yield for once to Rain's oft-repeated demand, which had been hitherto as regularly denied.

He took my flask, and with a guttural "How!" drained it at one gulp, without straining a muscle of his face. "Ugh! good! like Rain's heart," he remarked, as he handed the empty bottle to "Mac" with a self-satisfied look. Then, after a long pause, he joined in our hearty laughter, and added: "Wechasa Chischina (Little Man, as he always calls me) good! Potoshasha (Red Beard, his name for McFadden) good! Minnewaukan good! All heap good!"

"Something's come over the old man," laughed Mac. "His heart's good to-night. Suppose we take advantage of it, as the boys are all over at the Oglala tepees, and get Rain to turn his heart inside out. Here, give me my hat and I'll get the flask refilled and bring back Harry with me to interpret." Off he went like a shot, leaving me to entertain Rain as best I could with my small knowledge of the Sioux lingo and signs.

McFadden soon returned, bringing Harry McLaughlin, our interpreter. It didn't take long to get Rain started; after he had had another pull at the flask, he said:

"If you want a story, I will tell a true one. It's about myself. I was a bad man and dangerous to fool with before I had to walk with crutches. My heart's good now, but it was all the time bad when I was a fighter and

RAIN-IN-THE-FACE

From a sketch from life made in 1894 by Edward Esmonde

a hunter. The maidens admired me, but the bucks were afraid of me. I would rather fight than eat. The long swords (soldiers) trembled when they knew I was near, and the Rees and Crows always felt of their hair every morning to see if their scalps were still on when Rain was near by."

Here Harry headed him off, for it's natural for an Indian to boast, and if any one will listen he will sing his own praises for hours at a stretch.

"Yes, we know you were a bad man and a fighter from 'way back," said Harry, "but we want to hear about the time Tom Custer made you take water. If you were such a brave man, how did it happen that a little man like Tom Custer got the best of you?"

This had the desired effect, and Rain winced under such a reflection on his bravery, for he measured forty-six inches around his chest, stood five feet nine inches, and weighed about 195 pounds at that time, while Captain Tom Custer was under the average weight and height.

With great deliberation and much gesticulation, Rain told his version of the incident in question, and much to our surprise he continued on and related his version of the "Custer Massacre."

Now, since nearly all the officers in the Regular Army, as well as all the agents of the Interior Department, have failed to get him even to speak of this fight (their trying, coaxing, and threatening for years has been in vain), and since Rain gets the credit of being the slayer of Custer, and has been immortalized in verse by Longfellow, it was a pleasant surprise to have this unexpected revelation. I am writing it down as nearly like McLaughlin interpreted it as I can.

"Two years before the big fight," he said, "Gall and

Sitting Bull had their camp at Standing Rock. All were hostiles. They were Unkpapa Sioux and fighters who never feared an enemy any more than a buffalo calf. The Yanktonais (friendly Indians) were coffee coolers (cowards) and hung around the agency which was at Lincoln then (Fort Abraham Lincoln on the upper Missouri now). We used to have great times in the hostile camp, dancing, running races, shooting and playing games. Buffalo and deer were plenty, and we had many ponies. I was a great fellow with the girls. They used to tease me to get me mad — when I got mad I knew no reason, I wanted to fight. One night a girl dared me to go up to Fort Lincoln and kill a white man. I told her it was too risky, as the long swords always kept watch. Besides, the Rees (another tribe of Indians, employed by the government as scouts) had their lodges on the hill back of the fort. The wood-choppers were camped between the fort and the river. She said:

"'A brave man fears nothing. If you are a coward, don't go. I'll ask some other young man who isn't afraid, if he hasn't danced in the Sun Dance.' (This was a torture dance in which Rain-in-the-Face subsequently underwent the most horrible self-torture ever inflicted.*)

* The Sun Dance is that ceremonial performance in which the young Sioux aspirant gives that final proof of endurance and courage which entitles him to the *toga virilis* of a full-fledged warrior. One feature of it is the suspension in air of the candidate by a rawhide rope passed through slits cut in the breast, or elsewhere, until the flesh tears and he falls to the ground. If he faints, falters, or fails, or even gives way momentarily to his anguish during the period of suspension, he is damned forever after, and is called and treated as a squaw for the rest of his miserable life.

Rain-in-the-Face was lucky when he was so tied up. The tendons gave way easily, and he was released after so short a suspension that it was felt he had not fairly won his spurs. Sitting Bull, the chief medicine man, decided that the test was unsatisfactory. Rain-in-the-Face thereupon defied Sitting Bull to do his worst, declaring there was no test which could wring a murmur of pain from his lips.

Sitting Bull was equal to the occasion. He cut deep slits in the back over the kidneys — the hollows remaining were big enough almost to take in a closed fist years after —

"The other girls laughed, but the young men who heard it didn't. They feared me. I would have killed them for laughing. I went to my lodge and painted sapa (black, the color used when they go on the war-path), took my gun, my bow, my pony. Sitting Bull had forbidden any one to leave camp without his permission. I skipped off under cover of darkness and went up to Lincoln (forty-five miles north, opposite the present site of Bismarck, North Dakota). I hung around for two days, watching for a chance to shoot a long sword. I had plenty of chances to kill a Ree squaw, plenty to kill a wood-chopper, but I wanted to carry back the brass buttons of a long sword to the girl who laughed at me. I did so, and she sewed them on her shawl. One morning I saw the sutler (store keeper) and a horse medicine man (U. S. Veterinary Surgeon Hontzinger) go out to a spring; Long Yellow Hair (General Custer) and his men were riding back about 100 yards. I rushed up and shot the sutler and brained the horse medicine man with my war club; then I shot them full of arrows and cut off some buttons. Long Yellow Hair heard the shot and his troop charged back. I didn't have time to scalp the men I got. I jumped on my pony and yelled at them

and passed the rawhide rope through them. For two days the young Indian hung suspended, taunting his torturers, jeering at them, defying them to do their worst, while singing his war songs and boasting of his deeds. The tough flesh muscles and tendons would not tear loose, although he kicked and struggled violently to get free. Finally, Sitting Bull, satisfied that Rain-in-the-Face's courage and endurance were above proof, ordered buffalo skulls to be tied to his legs, and the added weight with some more vigorous kicking enabled the Indian stoic to break free. It was one of the most wonderful exhibitions of stoicism, endurance, and courage ever witnessed among the Sioux, where these qualities were not infrequent. Rain-in-the-Face had passed the test. No one thereafter questioned his courage. He was an approved warrior, indeed. It was while suspended thus that he boasted of the murder of Dr. Hontzinger, and was overheard by Charlie Reynolds, the scout, who told Custer and the regiment.

Mr. Edward Esmonde, a companion of Mr. Thomas during the season he had Rain-in-the-Face and his fellows at the World's Fair in Chicago and afterward at Coney Island in his charge, gave me the information in this note. — C. T. B.

to catch me. They chased me to the Cannon Ball (a small river twenty-five miles south). Charlie Reynolds (a scout) knew me and told Long Yellow Hair who did this brave deed.

"Next winter I went to the agency store at Standing Rock. I drew no rations — I hadn't signed the paper. (All the Indians who signed a peace treaty and consented to live on the reservation, under military orders, were furnished with rations by the government twice a month. The hostiles had to live by hunting.) Istokscha (One Arm, the Hon. H. S. Parkin) was running the store then. I wasn't afraid of any of them. Little Hair (Capt. Tom Custer) had thirty long swords there. He slipped up behind me like a squaw, when my back was turned. They all piled on me at once; they threw me in a sick wagon (ambulance) and held me down till they got me to the guard-room at Lincoln.

"I was treated like a squaw, not a chief. They put me in a room, chained me, gave me only one blanket. The snow blew through the cracks and on to me all winter. It was cold. Once Little Hair let me out and the long swords told me to run. I knew they wanted to shoot me in the back. I told Little Hair that I would get away some time; I wasn't ready then; when I did, I would cut his heart out and eat it. I was chained to a white man. One night we got away. They fired at us, but we ran and hid on the bank of Hart river in the brush. The white man cut the chains with a knife (a file). They caught him next day.

"I rejoined Sitting Bull and Gall. They were afraid to come and get me there. I sent Little Hair a picture, on a piece of buffalo skin, of a bloody heart. He knew I didn't forget my vow. The next time I saw Little Hair, ugh! I got his heart. I have said all."

And, Indian-like, he stopped.

We all knew that the greater part of this was true, since it tallied with the government account of the death of the sutler and Surgeon Hontzinger. But we wanted to hear how he took Tom Custer's heart. McFadden, who is quite an artist as well as an actor of note, had made an imaginary sketch of "Custer's Last Charge." He got it and handed it to Rain, saying:

"Does that look anything like the fight?"

Rain studied it a long time, and then burst out laughing.

"No," he said, "this picture is a lie. These long swords have swords—they never fought us with swords, but with guns and revolvers. These men are on ponies—they fought us on foot, and every fourth man held the others' horses. That's always their way of fighting. We tie ourselves onto our ponies and fight in a circle. These people are not dressed as we dress in a fight. They look like agency Indians—we strip naked and have ourselves and our ponies painted. This picture gives us bows and arrows. We were better armed than the long swords. Their guns wouldn't shoot but once—the thing wouldn't throw out the empty cartridge shells. (In this he was historically correct, as dozens of guns were picked up on the battle-field by General Gibbon's command two days after with the shells still sticking in them, showing that the ejector wouldn't work.) When we found they could not shoot we saved our bullets by knocking the long swords over with our war clubs—it was just like killing sheep. Some of them got on their knees and begged; we spared none—ugh! This picture is like all the white man's pictures of Indians, a lie. I will show you how it looked."

Then turning it over, he pulled out a stump of a lead

pencil from his pouch and drew a large shape of a letter S, turned sideways.* "Here," said he, is the Little Big Horn river; we had our lodges along the banks in the shape of a bent bow."

"How many lodges did you have?" asked Harry.

"Oh, many, many times ten. We were like blades of grass. (It is estimated that there were between four and six thousand Indians, hence there must have been at least a thousand lodges.)

"Sitting Bull had made big medicine way off on a hill. He came in with it; he had it in a bag on a coup-stick. He made a big speech and said that Waukontonka (the Great Spirit) has come to him riding on an eagle. Waukontonka had told him that the long swords were coming, but the Indians would wipe them off the face of the earth. His speech made our hearts glad. Next day our runners came in and told us the long swords were coming. Sitting Bull had the squaws put up empty death lodges along the bend of the river to fool the Ree scouts when they came up and looked down over the bluffs. The brush and the bend hid our lodges. Then Sitting Bull went away to make more medicine and didn't come back till the fight was over.

"Gall was head chief; Crazy Horse led the Cheyennes; Goose the Bannocks. I was not a head chief — my brother Iron Horn was — but I had a band of the worst Unkpapas; all of them had killed more enemies than they had fingers and toes. When the long swords came, we knew their ponies were tired out; we knew they were fooled by the death lodges. They thought we were but a handful.

* Rain-in-the-Face afterward drew a picture or map of the battle of the Little Big Horn, on the back of a handsome buckskin hunting-shirt. A cut of this picture appears on the following page. It is believed to be the only map of the battle drawn by one of the Indian participants therein.—C. T. B.

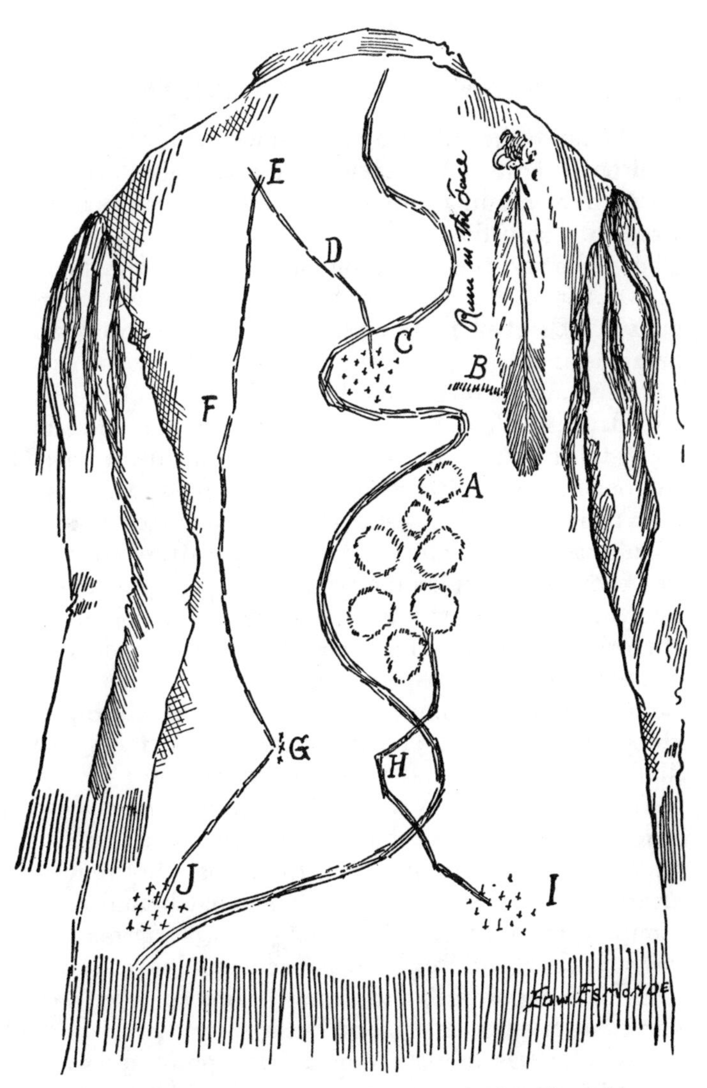

Key to map of Custer battle-field. Drawn on back of buckskin coat by Rain-in-the-Face for Edward Esmonde. Said to be the only map of the battle-field ever made by an Indian.

A, Camp of the Indians; B, Reno's Skirmish Line; C, Timber Where Reno's Horses Were; D, Reno's Retreat; E, Reno Joined Benteen; F, Custer's Trail; G, Custer's First Stand; H, Squaws and Children Crossed River; I, Where Squaws Went Into Camp After Re-crossing River; J, Where Last Stand Was Made and Custer Was Killed.

"We knew they made a mistake when they separated. Gall took most of the Indians up the river to come in between them and cut them off. When we saw the Ree scouts had stayed back with Long Yellow Hair, we were glad. We saw them trotting along, and let them come in over the bluffs. Some of our young men went up the gully which they had crossed and cut them off from behind.

"Then we showed our line in front, and the long swords charged. They reeled under our fire and started to fall back. Our young men behind them opened fire. Then we saw some officers talking and pointing. Don't know who they were, for they all looked alike. I didn't see Long Yellow Hair then or afterward. We heard the Rees singing their death song — they knew we had them. All dismounted, and every fourth man held the others' ponies. Then we closed all around them. We rushed like a wave does at the sand out there (the ocean beach) and shot the pony holders and stampeded the ponies by waving our blankets in their faces. Our squaws caught them, for they were tired out.

"I had sung the war song, I had smelt the powder smoke. My heart was bad — I was like one that has no mind. I rushed in and took their flag; my pony fell dead as I took it. I cut the thong that bound me. I jumped up and brained the long sword flag-man with my war club, and ran back to our line with the flag.

"The long sword's blood and brains splashed in my face. It felt hot, and blood ran in my mouth. I could taste it. I was mad. I got a fresh pony and rushed back, shooting, cutting, and slashing. This pony was shot, and I got another.

"This time I saw Little Hair. I remembered my vow. I was crazy; I feared nothing. I knew nothing would hurt me, for I had my white weasel-tail charm on.* (He wears the charm to this day.) I don't know how many I killed trying to get at him. He knew me. I laughed at him and yelled at him. I saw his mouth move, but there was so much noise I couldn't hear his voice. He was afraid. When I got near enough I shot him with my revolver. My gun was gone, I don't know where. I leaped from my pony and cut out his heart and bit a piece out of it and spit it in his face. I got back on my pony and rode off shaking it. I was satisfied and sick of fighting; I didn't scalp him.

"I didn't go back on the field after that. The squaws came up afterward and killed the wounded, cut their boot legs off for moccasin soles, and took their money, watches, and rings. They cut their fingers off to get them quicker. They hunted for Long Yellow Hair to scalp him, but could not find him. He didn't wear his fort clothes (uniform), his hair had been cut off, and the Indians didn't know him. (This corroborates what Mrs. Custer says about her husband's having his long yellow

* Notwithstanding his "white weasel-tail charm," Rain-in-the-Face was wounded in this battle. A bullet pierced his right leg above the knee. Among the plunder which fell to him after the action was over was a razor taken from the person of some dead soldier. With this razor the wounded man essayed some home-made surgery. First he cut deeply into the front of his leg, but failed to reach the bullet. Then he reached around to the back of his leg and chopped recklessly into the flesh from that quarter. He got the bullet, also several tendons, and narrowly missed cutting the artery and bleeding to death. He was lame and had to walk on crutches all his life thereafter.

—Statement of Mr. Esmonde.

Colonel Godfrey, in his *Century* article, relates a similar instance of courage and endurance on the part of one of his troopers:

"Among the wounded was Saddler 'Mike Madden,' of my troop, whom I promoted to be sergeant on the field for gallantry. Madden was very fond of his grog. His long abstinence had given him a famous thirst. It was necessary to amputate his leg, which was done without administering any anæsthetic; but after the amputation, the surgeon gave him a good stiff drink of brandy. Madden eagerly gulped it down, and his eyes fairly danced as he smacked his lips and said:

"'M-eh, doctor, cut off my other leg.'" — C. T. B.

curls cut at St. Paul some weeks before he was killed.)

"That night we had a big feast and the scalp dance. Then Sitting Bull came up and made another speech. He said: 'I told you how it would be. I made great medicine. My medicine warmed your hearts and made you brave.'

"He talked a long time. All the Indians gave him the credit of winning the fight because his medicine won it. But he wasn't in the fight. Gall got mad at Sitting Bull that night. Gall said: 'We did the fighting, you only made medicine.' It would have been the same anyway. Their hearts were bad toward each other after that, always.

"After that fight we could have killed all the others on the hill (Reno's command) but for the quarrel between Gall and Sitting Bull. Both wanted to be head chief. Some of the Indians said Gall was right and went with him. Some said Sitting Bull was. I didn't care, I was my own chief and had my bad young men; we would not obey either of them unless we wanted to, and they feared us.

"I was sick of fighting—I had had enough. I wanted to dance. We heard more long swords were coming with wheel guns (artillery, Gatlings). We moved camp north. They followed many days till we crossed the line. I stayed over there till Sitting Bull came back, and I came back with him. That's all there is to tell. I never told it to white men before."

When he had finished, I said to him: "Rain, if you didn't kill Long Yellow Hair, who did?"

"I don't know. No one knows. It was like running in the dark."

"Well," asked Mac, "why was it Long Yellow Hair

wasn't scalped, when every one else was? Did you consider him too brave to be scalped?"

"No, no one is too brave to be scalped; that wouldn't make any difference. The squaws wondered afterward why they couldn't find him. He must have laid under some other dead bodies. I didn't know, till I heard it long afterward from the whites, that he wasn't scalped."

"How many Indians were killed in the fight?"

"I don't remember, but about ten and four or ten and six."

"How about Curley, the Crow scout, who claims to have escaped?" asked Mac.

"Ugh! I know Curley. He is a liar. He never was in the fight. His pony stumbled and broke something. He stayed behind to fix it. When he heard the firing, he ran off like a whipped dog. One long sword escaped, though; his pony ran off with him and went past our lodges. They told me about it at Chicago. I saw the man there, and I remembered hearing the squaws tell about it after the fight."

Rain-in-the-Face (Itiomagaju) is about sixty years of age now, and is the only chief that survives to tell the tale of the Custer fight. Gall and Sitting Bull have both gone to hunt the white buffalo long since. Rain can write his name in English. I taught him to do it at the World's Fair in order to sell Longfellow's poem, entitled "The Revenge of Rain-in-the-Face." He doesn't know the significance of it after he writes it. His knowledge of English is confined to about thirty words, but he can't say them so any one can understand him, though he can understand almost anything that is said in English. Like all other Indians, his gratitude is for favors to come and not for favors already shown. He is utterly heartless and unprincipled, physically brave but morally a coward. His redeeming feature lies in the fact that you can depend upon any promise he makes, but it takes a world of patience to get him to promise anything. Even at the age of sixty he is still a Hercules. In form and face he is the most pronounced type of the ideal Fenimore Cooper, dime novel Indian in America.

CHAPTER EIGHT

Two Interesting Affairs

I. The Fight on the War Bonnet

BEFORE entering upon a detailed description of the larger events of the campaign after the Battle of the Rosebud and Little Big Horn, two smaller affairs are worthy of mention. One, though nothing but a skirmish, was of great importance in determining the final result. The other well illustrates something of the adventurous life and perilous duty of a soldier in Indian warfare.

On Saturday, July 15, 1876, the Fifth Cavalry, under General Wesley Merritt, was marching toward Fort Laramie, under orders to join Crook. At noon word was received from the agency that a body of Cheyennes, numbering, perhaps, one thousand warriors, who had heretofore remained quiet on the reservation at the Red Cloud Agency, on the White River, South Dakota — the Pine Ridge Agency — was about to break away and join the Indians in the field. Their minds had been inflamed by the story of Crook's defeat and the account of the disaster to the Seventh Cavalry. They thought they saw unlimited opportunities for plunder, scalp-taking, and successful fighting — therefore they decided to go on the war-path without delay. There were not troops enough near the agency to prevent this action, which was entirely unsuspected anyway.

The orders for Merritt to join Crook were imperative; but, in view of this news, the general decided to disregard them for the present. He realized that he could perform no better service than heading off this body of Cheyennes, and either defeating and scattering them or, better still, forcing them back to the agency.

The trail they would have to take would cross a creek in the extreme southeast corner of South Dakota, called the War Bonnet,* some eighty-five miles, by the only practicable route, from where the Fifth Cavalry then was. The Indians were a much shorter distance from it. Merritt would have had to march around, practically, three sides of a square, owing to the configuration of the country, to reach that point, which was the best place for miles around, within the knowledge of W. F. Cody (Buffalo Bill), his chief scout, to intercept the flying Cheyennes.

Merritt did not hesitate an instant after learning the news. He put his command in motion immediately, and by a forced march of thirty-one hours, got to the crossing in good time. There was no evidence that the Cheyennes had passed. The troopers concealed themselves in ravines under the bluffs, and waited for the Indians.

Early on the morning of July 17th, the pickets, commanded by Lieutenant Charles King,† observed the approach of the Indians. At about the same time Merritt's wagon train, under Lieutenant Hall, with two hundred infantrymen spoiling for a fight, concealed in the wagons as a guard, was observed toiling along, some four

* The frontiersmen translate this to "Hat Creek"; and that is the name it bears to-day — more's the pity!

† Afterward brigadier-general of volunteers in the Spanish-American War, and the author of many fascinating romances of army life.

miles to the southwest, in an endeavor to reach the rendezvous on the War Bonnet. The regiment remained carefully concealed, and the Indians, in high glee, thought they had the train at their mercy.

So soon as he sighted the Cheyennes, Lieutenant Hall despatched two troopers of his small cavalry escort ahead to the crossing to apprise Merritt that the Indians were at hand. An advance party of Cheyennes, superbly mounted and led by a gorgeous young chief, determined to intercept these troopers, who were ignorant of their peril. The two soldiers came down one trail which led through a ravine, the Indians came up another which led through another ravine. The troopers and the Cheyennes were hidden from each other, but both were in plain view of the picket on the hill. The two trails joined at the foot of the hill. The plain back of the wagon-train was black — or red, rather — with Indians coming up rapidly, although they were not yet near enough to attack.

Merritt and one or two other officers, with Buffalo Bill and a few of his scouts and several troopers, joined King on the hill. The main body of the Indians was too far away to attack, so the little advance party determined to wait until the Cheyennes, who were endeavoring to cut off the two soldiers, were close at hand and then fall upon them. Everybody withdrew from the crest of the hill except Lieutenant King, who was to give the signal, when the party below should sally around it and fall on the Cheyennes.

King, who has described the situation with masterly skill in his "Campaigning with Crook," flattened himself out on the brow of the hill, with nothing showing but the top of his hatless head and his field glass, and watched the soldiers rapidly galloping up one trail and the Indians more rapidly rushing down the other. He waited until the Indians had almost reached the junction. Then he gave the signal. Merritt's escort and Cody's scouts raced around the base of the hill, and dashed slap into the faces of the astonished Cheyennes. Two Indian saddles were emptied in the twinkling of an eye. Such was the impetus of their charge that the Indians scarcely had time to rein in their steeds before the white men were upon them.

Buffalo Bill shot the leader of the war party, a famous young chief named Yellow Hand, through the leg. The bullet also pierced the heart of the pony Yellow Hand was riding. Both crashed to the earth. In spite of his pain, Yellow Hand dragged himself to his feet and fired at the scout, killing his horse. The two, not twenty paces apart, exchanged shots the next instant. The Indian missed, but Buffalo Bill sent a bullet through Yellow Hand's breast. The Indian reeled, but before he fell Cody leaped upon him and drove his knife into his gallant enemy's heart. Yellow Hand was a dead Indian when he struck the ground. "Jerking the war bonnet off," he says, "I scientifically scalped him in about five seconds."*

Yellow Hand had recognized Buffalo Bill, and had virtually challenged him to this duel. "The first scalp for Custer!" shouted Cody, waving his trophy in the air.

Some of the other Indians had now come within range. They opened fire upon the little party; the bullets zipped around them in every direction, one narrowly grazing General Merritt. They nicked a horse here and there, but, as usual, their marksmanship was execrable.

As the little party charged the Indians, Merritt had

*"The Adventures of Buffalo Bill." By Colonel William F. Cody. Harper & Brothers, 1904.

directed King to order the rest of the regiment to advance. In the midst of the firing, the splendid troops of the dandy Fifth came bursting through the ravines and over the hills, making for the Cheyennes on the gallop. At the same time Lieutenant Hall's infantrymen scrambled out of their wagons and sent a few volleys at the Cheyennes at long range.

A more astonished body of Indians the United States has probably never contained. They hadn't the slightest idea that there was a soldier within five hundred miles, except those in the wagon train which they had expected to capture. They had anticipated no trouble whatever in joining Sitting Bull, and now they found themselves suddenly face to face with one of the finest cavalry regiments in the service. What were they to do? They hadn't much time to decide, for the cavalry were after them at full gallop. They turned and fled incontinently. They stood not on the order of their going, but went at once.

If they could get back to the reservation, they would be free from attack. They fled at the highest possible speed of their horses, throwing aside everything they possessed, save their guns and ammunition, in their frantic desire to get away. For thirty miles Merritt and his men pursued them with the best will in the world to come up with them; but the horses of the soldiers were more or less tired from their long march of the day before, and the Indians, lightly equipped and on fresh horses, finally succeeded in escaping. By nightfall the whole party was back on the reservation. Thereafter care was taken that they found no further opportunity to go on the war-path.

The coöperation of this splendid body of Indians with that under the command of Crazy Horse might possibly have turned the scale in some of the hotly contested battles, and Merritt's promptness was greatly commended by the authorities. Buffalo Bill received the chief glory of the little adventure from his dramatic duel with Yellow Hand, in full view of soldiers and Indians.

II. The Sibley Scout

The other event is known in army records as The Sibley Scout. While General Crook was waiting for reinforcements and additional supplies at his camp on Goose Creek, near the Tongue River, he decided to send out a scouting party to see what had become of his friend, Crazy Horse, who had handled him so severely at the Rosebud a few weeks before.

Lieutenant Frederick W. Sibley, of E Troop, of the Third Cavalry, an enterprising but cool-headed young officer, was given command of twenty-five picked men from the regiment. With him went scouts Frank Gruard and Baptiste Pourier, commonly known as "Big Bat," to distinguish him from another scout, Baptiste, a smaller man. To the party also were attached John Becker, mule packer, and the indefatigable Finerty, the war correspondent of the Chicago *Times*, making a total of thirty men.

Each man carried one hundred rounds of ammunition on his person, and a few days' rations in his saddle-bags. They started on the 6th of July. On the 7th they had reached the Rosebud, some fifty miles away from Crook's camp. There they came across the Indians. Gruard and Pourier observed them from the top of a hill, behind which the rest of the expedition halted. There were hundreds of them, apparently, and the scouts rejoined the command immediately. To take the back track was impossible. Therefore, they struck westward over the mountains, leading their horses. The Indians, marching slowly southward, soon came upon the trail of the party, and followed it at some distance. Urged by the imminence of their peril, the men, led by the unerring Gruard, who was familiar with all the ramifications of the Big Horn Range, since he had often hunted there during his captivity with the Sioux, did some rapid mountain climbing, and finally thought they had escaped pursuit, especially as no one could ride up the trail up which they had climbed, and these Indians were poor trailers when on foot. Having progressed some five miles over terrific trails, they halted in a little glade under the shade of some trees, unsaddled their horses, made coffee, and ate dinner. Feeling themselves safe from pursuit, they rested for several hours, and it was not until late in the afternoon that they took up their march again.

The going here was easier than before, and they could mount their horses once more. Presently they trotted into a level, thickly wooded valley. The trail led along the right side of the mountain, which was broken and rugged. There were woods to the left and in front of them, and high rocks and open timber on the right. John Becker, who brought up the rear, suddenly alarmed everybody by the shout of "Indians, Indians!"

The next instant the timber and boulders to the right were alive with a war party of Sioux and Cheyennes, not two hundred yards away — not the same party they had seen in the valley, by the way. So soon as the Indians appeared they opened fire. Again their shooting was bad. Not a trooper was hurt, although a number of horses were hit, some seriously. Sibley acted with prompt decision. A word with Gruard determined him in his course. Under a spattering fire from the Indians, the party turned to the left and raced for the thick timber as fast as they could go. They threw themselves to the ground in a semi-circular line so soon as they reached the woods, tied their horses to the trees back of them, and taking advantage of fallen logs and boulders as a breastwork, opened fire upon the Indians, who, on their part, sought concealment and commenced firing in earnest. The soldiers were well protected in the forest, however, and although the Indians killed many of the horses, they did not hit any of the troopers.

The party was now overwhelmingly outnumbered. There were already several hundred Indians engaged. Their leader was a magnificent young Cheyenne chief, dressed in a suit of white buckskin. It was afterward learned that his name was White Antelope. Gruard was recognized by the Indians, who were desirous of taking him alive. After firing for perhaps half an hour, White Antelope led the Indians on foot in a direct charge on the woods.

Sibley ordered his men to hold their fire until they could make every shot tell. They mowed the advancing Indians down in scores. White Antelope was seen to leap into the air and fall. He had been pierced, it was afterward learned, by several bullets, and started for the happy hunting grounds then and there. The charge was handsomely repulsed, and the Indians retired in confusion, although still keeping up a severe fire.

It was evident to every one that the Indians would hold the soldiers in play until they were joined by other war parties — indeed, their numbers were increased already — when Sibley's detachment would be surrounded and exterminated. Gruard, therefore, proposed abandoning the horses — most of them had been killed any-

CHIEF TWO MOON OF THE NORTHERN CHEYENNES

Allies of the Sioux at Little Big Horn

Painted from life by F. A. Burbank

way — and that the whole party should steal away through the timber and endeavor to escape over the mountains on foot. Firing two or three volleys and then keeping up a scattering fire for a short time to make the Indians think they were on the alert, the troopers, exercising the greatest caution, one by one crawled through the underbrush until they were hidden by the forest trees. Then everybody got up on his feet and ran like mad.

Gruard, whose instincts as a guide were of the highest order, led them over magnificent mountains, through gloomy cañons, past overhanging cliffs, along impossible trails on the sides of tremendous precipices, one of which stretched for several hundred feet below them and three hundred feet above them, almost sheer. Not being mountaineers, they would have been utterly unable to have followed the scout had it not been for the Red Terror that lurked behind. They had succeeded in getting, perhaps, a mile away from and some distance above the valley, when they heard several heavy volleys, followed by a series of wild yells, which apprised them that the Indians had at last rushed their camp. They were so confident of escape now that they actually burst into roars of laughter at the thought of the Indian disappointment when the attackers found their victims had decamped. Those Indians were not accustomed to hunt on foot. An Indian off a horse is about as awkward as a sailor on one. The pursuit was soon abandoned, and the soldiers left to follow their course unmolested. Theirs had been a lucky escape. Without Gruard, they had all been killed.

The day was frightfully hot. The fast going caused by the exigencies of the occasion and the desperate nature of the climbing increased their discomfort. The men threw away everything in the way of superfluous clothing which would impede their progress or tire them in their hurry, save their weapons and ammunition. They camped that night, or halted, rather, for there was nothing with which to camp, on the crest of the range. It turned very cold, a terrible storm arose, and they suffered severely. They had nothing to eat; their provisions had been in their saddle-bags, and they had not dared to take them in their attempt to escape, lest the suspicions of the Indians should be excited by their efforts.

The next day, the 9th of July, they started down the mountain. Gruard's instincts were not at fault. He led them to the foot-hills overlooking Crook's main camp far away. In order to reach the plain they had to cross a rapid mountain brook, the water of which came almost up to their necks. Two men who could not swim and who were in a very nervous condition from their exciting adventures, stubbornly refused to try to cross the stream, even with the assistance of their comrades. They chose rather to hide themselves where they were, and begged that help might be sent back to them. The rest of the party managed to cross and started for the camp, still about fifteen miles distant. They were met in the evening by a scouting party of soldiers, who brought them back to camp.

Their clothing and shoes were torn to ribbons, and they were greatly exhausted from the terrible strains and hardships to which they had been subjected. That they escaped at all was a miracle, due to the coolness of young Sibley and the marvelous skill of Gruard. A detachment went back for the two men who had remained behind and brought them back to the camp.

Lieutenant Sibley reported to General Crook that he

had found the Indians, but whether that statement is accurate is a question. It would, perhaps, be more truthful to say that the Indians had found him. Sibley and Gruard were highly complimented by Crook; and Mr. Finerty, who had displayed great courage, wrote a graphic account of it, from which this brief sketch has been abridged.

CHAPTER NINE

The First Success

I. Crook and Mills at Slim Buttes

AFTER the defeat of General Custer, and the successful retreat of the Sioux and Cheyennes from the Little Big Horn, the government hurried reinforcements into the field, and ordered Crook and Terry to press the pursuit of the Indians with the greatest vigor. It was not, however, until nearly a year after the disaster on the Little Big Horn that the Sioux war was concluded, and it was not until after the Indians had met with several crushing defeats and had been pursued until they were utterly exhausted that peace was declared.

The greatest individual factor in bringing about this much desired result was General George Crook, a celebrated cavalryman during the Civil War, and a more celebrated Indian fighter after its close. With unwearied tenacity and vigor he pursued the savages, striking them through his subordinates whenever and wherever they could be found. The terrible persistence with which he urged his faint, starving, foot-sore, tattered soldiers along the trail, to which he clung with a resolution and determination that nothing could shake, entitles him to the respect and admiration of his countrymen—

a respect and admiration, by the way, which was fully accorded him by his gallant and equally desperate foes.

After Crook, the men who brought about the result were, first and foremost, Nelson A. Miles — singularly enough not a cavalryman, but the Colonel of the Fifth Infantry; and, next to him, Colonel Ranald S. Mackenzie, of the Fourth Cavalry, and Captain Anson Mills, of the Third, whom we have already noted doing gallant service at the Battle of the Rosebud. Miles had been ordered into the field to reinforce Terry's shattered and depleted column.

After much marching and scouting, the columns of Terry and Crook combined; but Terry's forces were in bad condition, and his command was soon withdrawn from the field. What was left of the Seventh Cavalry was sent back to Fort Lincoln, whence they had started out with such bright hopes a few months before. Gibbon's command was returned to Montana, where it had been made up, on account of the threatening aspect of things in that quarter, and Terry retired from active campaigning to resume command of his department. Miles, as we shall see, was sent to the Yellowstone.

Crook was left alone in the active pursuit. Space and time are lacking to describe the details of the wonderful marches he made on the trails of the Indians — now under burning suns, which parched the ground until it was as bare as the palm of a hand; again through torrents of drenching rains, which succeeded the fierce heat; and, finally, through the snows and cold of a winter of unexampled severity. During the summer there was no forage for the horses of the cavalry nor for the very small pack train, and rations for the men became shorter and shorter. Finally, early in September, the supply of provisions was reduced to two and a half days' rations. Crook calculated that they could march two weeks on that amount.

They supplemented the rations by living on horse and mule meat and a few wild onions which they could gather from time to time in spots which had escaped the universal baking of the summer. At last the command literally reached the end of its resources. The Indians were in bad condition, too, but their situation was not nearly so desperate as was that of Crook and his men. The Indians were worn out and exhausted by the energetic and relentless pursuit which had been hurled after them by the indomitable commander, but they still had plenty to eat, and they had managed to keep ahead of him, and to avoid various scouting columns.

On the 7th of September, 1876, Crook realized that his men had reached the limit of their endurance, and that forage and food must be procured or they would all die in the wilderness. The Indians had swept the country bare of game, and the sun had swept it clean of fodder. One hundred and fifty of the best men — that is, those who showed the fewest signs of the hardships they had undergone — with the best horses and the last of the mules, were formed into an advance party under Captain Anson Mills, of the Third Cavalry. Mills had instructions to push on to Deadwood City, one of the new towns in the Black Hills, to get provisions, "Any kind of provisions, for God's sake!" which he could bring back to the rest of the army, now in a destitute condition.

Mills was not expected to hunt for, or to fight, Indians — primarily, that is. He was to go for food in order to keep the army from starving; but as he marched southward, his scouts discovered a large village of forty or fifty lodges at a place called Slim Buttes, in the northwest corner of South Dakota. The tepees had been pitched on a little rising from the banks of a small stream called Rabbit Creek. The place was inclosed on three sides by a series of tall cliffs, whose broken sides seemed here and there to have been cut in half-formed terraces, making the ascent easy. Little ravines and small cañons ran through the buttes, gradually ascending until they met the plateau on top.

Mills instantly determined to attack the camp — a wise and soldierly action on his part. He made his dispositions with care. Reaching the vicinity of the camp, he halted in a deep gorge on the night of September 8, and prepared for battle early the next morning. The night was dark, cold, and very rainy, and the tired men suffered greatly. Marching out at dawn, Mills succeeded at daybreak in surprising the camp, which proved to be that of a band of Sioux led by American Horse, one of the most prominent chiefs. Leaving Lieutenant Bubb with the pack train and the lead horses, Mills directed Lieutenant Schwatka, afterward so well known from his Arctic explorations, to charge directly into the village with twenty-five mounted men. The remainder of his force he dismounted and divided into two parties, under Lieutenants Von Luettwitz and Crawford, respectively, with orders to move on the camp from different sides.

The attack was a complete success. The village was taken with but little loss. Some of the Sioux were killed and others captured, but most escaped through the ravines to the plateau surrounding the valley. One heroic but unfortunate little band, consisting of American Horse and four warriors, with fifteen women and children, was driven into one of the cañons which ended in a cave. One or two of the soldiers had been wounded in the attack. Lieutenant Von Luettwitz, who had fought all through the Franco-Austrian War in Italy, and who was a veteran of the Civil War, was shot in the knee and so badly wounded that his leg had to be amputated on the field.

Being now in complete command of the village, the pack train was ordered up and the captured village was examined. To the joy of Mills and his soldiers, an immense quantity of provisions, in the shape of meat, forage, and other stores, was discovered.

There still remained the little band of savages in the ravine to be dealt with. A detachment was ordered to drive them out. The Indians had been busy making rifle-pits, and as the soldiers advanced to storm the cave, they were met with a rapid and well-directed fire. Two of them were shot dead and others wounded.* The Indian position appeared to be impregnable. An interpreter crept near enough under cover to make himself heard, and asked their surrender. They replied to his command with taunts and jeers. They incautiously informed him, however, that Crazy Horse with his warriors was in the vicinity, and on being apprised of their situation by some of the fugitives, he would undoubtedly come to their rescue. Crazy Horse could have made short work of Mills and his hundred and fifty. Meanwhile the survivors of the village, which had contained a hundred warriors, formed an extended line on the buttes and opened fire on the soldiers.

* One of the scouts killed in this battle was a great admirer of Buffalo Bill, whose manners, methods, and appearance he aped as well as he could. He rejoiced in an unfortunate sobriquet, which was received in this wise: General Sheridan, seeking Buffalo Bill to lead a hunting expedition on one occasion, was met by this swaggerer, with the remark that Buffalo Bill was gone away, and when Buffalo Bill was gone he was Buffalo Bill himself. "The h—l you are!" said Sheridan contemptuously. "Buffalo Chip, you mean!" The poor braggart never got away from the name of "Buffalo Chip Charlie." He was a brave man for all his vanity, and the soldiers were sorry enough for their mockery when they buried him that night at the foot of the buttes, where he had fallen in the attack on the cave.

Mills acted promptly. He despatched a courier to Crook on the best horse in the command, to report the situation and ask him for reinforcements at once. Incidentally, he mentioned that a great quantity of provisions had been found. Then he made preparations to hold the place, and at the same time to prosecute his attack against the cave, all the time keeping up a smart fight with the men on the buttes. So soon as Crook received the message, he started forward, intending to take with him a select body of men; but the whole army, spoiling for a fight and hungry for a square meal, insisted on going along. They made a forced march, and reached Mills about half after eleven in the morning.

Crook immediately proceeded to dislodge the Indians in the cave. The men were led forward under a galling fire, to which the general, in spite of the entreaties of his staff, exposed himself with indifference. When they got in a position to command the cave, Crook, willing to spare his brave foemen, again asked them to surrender. His request was met by a decided negative. The men opened fire, and searched every cranny and recess of the cave with a storm of bullets. Gruard, one of the scouts, taking advantage of cover, crept to the very mouth of the cave, remained there unobserved, watched his opportunity, seized a squaw who incautiously exposed herself, and with her as a shield dashed forward and shot one of the warriors, escaping in safety himself.

II. The Death of American Horse

After two hours of firing, the death-chants of the squaws induced Crook to order a cessation for another parley. This time his request that the Indians surrender met with some response; for the squaws and children, to the number of thirteen, came reluctantly forth on his positive assurance that they would be protected. The braves refused to give up. They were confident that Crazy Horse would succor them. The engagement at once began again, but after it had lasted some little time the fire of the Indians ceased.

The offer of mercy was made a fourth time. A young Indian stepped out and received additional assurance that no harm should come if they surrendered. He went back into the cave and presently reappeared with another young warrior, supporting between them the tall, splendid figure of brave old American Horse. He had been shot through the bowels, and his intestines protruded from the wound. He was suffering frightful agony, and was biting hard upon a piece of wood to control himself. He handed his gun to Crook and gave up the contest. The surgeons with the command did everything they possibly could for him, but his wound was beyond human skill. That night, surrounded by his wives and children, he died, as stoically and as bravely as he had lived.

Inside the cave the rocky walls were cut and scored by the rain of bullets which had been poured into it, and lying on the floor were the bodies of the two Indian warriors, together with a woman and a child, who had been killed. The soldiers had not known, until the squaws came out, that there were any women or children there. The little band had sold their lives dearly. Even the women had used guns, and had displayed all the bravery and courage of the Sioux.

Too late Crazy Horse, with some six hundred warriors, appeared on the scene. Imagining he had only to deal with Mills' small force, he galloped gallantly forward to the attack at about five o'clock. He was greatly astonished at the number of antagonists developed thereby. He retired to the top of the buttes, and the soldiers in gallant style dashed after him. They scaled the cliffs, finally gaining the level plateau. Crazy Horse made one or two attempts to break through the line, but it was impossible, and seeing himself greatly outnumbered, he wisely retired, having sustained some loss.

The battle was one of the most picturesque ever fought in the West. Crook and his officers stood in the camp, the center of a vast amphitheater ringed with fire, up the sides of which the soldiers steadily climbed to get at the Indians, silhouetted in all their war finery against the sky. The loss of life on either side was not great, but the capture of the village and the provisions which had been accumulated for the winter was a serious one.

In the camp were discovered many articles that had belonged to the Seventh Cavalry — a guidon, money, one of Captain Keogh's gauntlets, marked with his name, orderly books, saddles, etc. Among other things, were letters written by officers and soldiers to friends in the East, some of them still sealed and ready for mailing. They must have come like voices from the dead when they reached those to whom they had been written.

CHAPTER TEN

A Decisive Blow

I. Mackenzie's Winter Battle

CROOK now gave over the pursuit, and returned to Fort Fetterman to organize a winter campaign. This expedition was one of the best equipped that ever started on an Indian campaign. It contained all arms of the service, with an abundance of everything necessary to success. To follow its marches to the Big Horn Range would reveal little of interest; but late in November it was learned, from a captured Cheyenne, that the principal Cheyenne village was located in a cañon through which flowed one of the main sources of Crazy Woman's Fork of the Powder River. Colonel Ranald S. Mackenzie was ordered, with the Indian scouts and ten troops of cavalry from the Second, Fourth, and Fifth regiments, to find and destroy the village.

The Cheyennes were not so numerous as the Sioux, and the greater number of their allies has sometimes caused people to minimize the quality of the Cheyennes; but no braver, more magnificent fighters ever lived than this same tribe. They had some of the Homeric qualities of the ancient Greeks. I believe it will generally be admitted that they were the finest of the Plains Indians. They were foemen worthy Mackenzie's or anybody else's steel. The battle which ensued was in some respects one of the most terrible in Western history, and in its results exemplified, as few others have done, the horrible character of the war. It was, perhaps, as great a contribution to the downfall of the Sioux as any single incident that occurred.

Mackenzie's men left the main encampment on the 23d of November. The ground was covered with snow. The weather was arctic in its severity. The scouts and friendly Indians — Pawnees, Crows, Shoshones, the hereditary enemies of the Cheyennes, including certain Cheyennes also who had entered the service of the United States* — had located the camp in Willow Creek Cañon. Some of the Indians had kept the camp under observation while Mackenzie brought up his troops. He had seven hundred and fifty cavalrymen and three hundred and fifty Indians. Halting at the mouth of the cañon, which he reached on the night of the 24th, he resolved to await the still hours before the break of day the next morning before delivering his attack.

The cañon was a gloomy gorge in the Big Horn Mountains. A swift, ice-bound river rushed over the rocks between precipitous walls, which soared into the sky for perhaps a thousand feet on either side. Numberless icy brooks poured their contents into the main stream through lateral cañons scarcely less forbidding in their appearance than the main one, and which made the trail of the creek almost impossible. Here and there the cañon widened, and in one of these open

* It is a singular thing to note the looseness of the tie with which the members of the various tribes were bound. Frequently we find bands of the same tribe fighting for and against the United States on the same field. One of the most fruitful causes of the success of our arms has been this willingness on the part of the Indians to fight against their own people, of which the government has been quick to avail itself.

places the Cheyennes, under the leadership of Dull Knife, had pitched their camp. They fondly believed the place impregnable—as, indeed, with careful guarding it would have been. The greatest precaution was taken by Mackenzie to prevent his men from making any noise. They stood in ranks by their horses in the snow in that polar cold, waiting for the order for the advance. Presently the moon rose, flooding the recesses of the ravine with silvery light, which sparkled with dazzling brilliancy upon patches of snow here and there on the dark walls.

Mackenzie, calculating that day would be breaking just about the time he would reach the camp from his present position, at last gave order to take up the march. With what relief the benumbed troopers sprang to their saddles and urged their shivering horses forward, can scarcely be imagined by dwellers in peaceful lands around warm firesides. As they struggled up the cañon they could hear the sound of dancing and revelry in the Indian camp, faintly blown back to them by the night wind. They learned afterward that the Cheyennes had just returned from a successful raid on the Shoshones, and that the dance was in celebration of an important victory they had gained. They halted again, therefore, until all was silence, before they once more advanced. Day was beginning to break as they reached the village.

The sleeping Indians in the camp had not the slightest suspicion that the enemy was within a hundred miles. The troops, cheering and shouting, burst upon them like a winter storm. Indians, when not apprehensive of attack, invariably sleep naked. The Cheyennes had just time to seize rifles and cartridge belts, while the women caught hasty blankets about the children, when the soldiers were upon them. Indeed, so quick and sudden was the attack that some of the warriors could not get out of the tepees. With their knives they slashed the wigwams, and from these openings fired upon the soldiers as they galloped through the village. Many were shot dead where a few moments before they had slept in peace.

Most of the pony herd was captured, and the village in a short time was in possession of Mackenzie. The Cheyennes, though overwhelmed, were undismayed. They had retreated headlong up the cañon, but were soon rallied by their subchiefs. Dull Knife, their leader, was found in the village with half a dozen bullets in him. He had fought gallantly in the open until he died.

Presently the Indians came swarming back along the side of the cañon. They occupied points of vantage, and, naked though they were in the frightful weather, with the thermometer ranging from ten to twenty degrees below zero during this campaign, they opened fire upon their opponents. Unless they could be dislodged, Mackenzie's position was untenable. He sent his Shoshone and other Indian scouts, who, animated with bitter hatred of the Cheyennes, were eager to obey his commands, to the summits of the cliffs to clear the Indians from them.

Meanwhile he directed Lieutenant John A. McKinney, with his troop, to charge and drive the Indians from a rocky eminence where they were concentrating and from which they were pouring a hot fire upon the soldiers. McKinney's charge was entirely successful, for he drove the Cheyennes back until he was stopped by a ravine. Wheeling his men, he attempted to find a crossing, when he was fired upon by a flanking party of Indians and instantly killed, being hit no less than six times. Six of his troopers were wounded, and a number of horses were shot. The troop was thrown into confusion, and some of the men started to retreat. Mackenzie, observing the situation, immediately ordered Captain John M. Hamilton and Major G. A. Gordon to charge to the rescue. The charge was gallantly made and stubbornly resisted.

The fighting was hand to hand, of the fiercest description; and the Cheyennes, while keeping the rest of Mackenzie's forces engaged, began concentrating on these two troops, which had been joined by Captain Davis, with his men. There was no reserve; the cavalry were all in, and this detachment might have been wiped out had it not been for the success of the Shoshones and other Indians, who cleared the key to the position on the summit of the plateau above the cañon, and then came to the assistance of the sorely beset soldiers. Twenty Cheyennes were killed here and several of the soldiers.

Relieved in a measure by these two movements, although not altogether, for the Cheyennes with their superior knowledge of the topography of the country could not be entirely dislodged from their position, and kept up a fierce fire upon the soldiers all day long, to which he could make little reply, Mackenzie sent back word to Crook of his success, and meanwhile began the destruction of the village. All the winter supplies for over a thousand Indians were there. The Cheyennes were a forehanded, prosperous tribe of Indians, as Indians go, and the property destroyed was enormous.

II. The Sufferings of the Cheyennes

What must have been the despair of the surprised warriors, with their women and children, naked, shiver-

Courtesy of The Century Co.

MACKENZIE'S MEN IN DULL KNIFE'S VILLAGE

Drawing by Frederic Remington

ing in the hills, as they saw their belongings consumed by the flames! It was simply impossible for them to maintain their position during the night. They had to move away or die of cold. As it was, twelve little Indian babies froze to death that awful night. Many of the older men and women were kept alive only by having their hands and feet, and in the case of the children, their whole bodies, thrust into the warm bodies of the few ponies not captured by the soldiers, which had been disemboweled for the purpose.

There was no fighting on the 26th. The Cheyennes took up a strong position six miles farther up the cañon, from which Mackenzie could not dislodge them, and on the 27th he started on his return to the camp. Crook, who made a forced march night and day, with Colonel Dodge and the infantry, who came forward with astonishing speed in spite of storm and cold, met Mackenzie retiring just after he left the cañon, and the whole army returned to the encampment.

The subsequent sufferings of the Indians were frightful. Naturally, they repaired to Crazy Horse, expecting that he would succor them, feed them, and clothe them. The Sioux and the Cheyennes had been warm friends and allies, and had fought together on many a field. Had they come in their prosperity, Crazy Horse would have given them a warm welcome. As it was, he had little with which to support his own band during the winter, owing to Crook's pursuit of him, and with short-sighted, yet natural—from an Indian point of view—policy, he refused to receive these Cheyennes, or to share anything with them.

Exasperated beyond measure by their treatment by the Sioux, and swearing eternal vengeance upon Crazy Horse, the wretched band struggled into the nearest agency and surrendered, and in the following spring moved out with the soldiers against Crazy Horse and his men.

It is appalling to think of that night attack in that awful weather upon that sleeping camp—to read of those wretched women and children, wandering naked in that bitter cold; to learn of those little ones frozen to death; of the old men and women abandoned by the road to die—yet there is another side to the picture, scarcely less horrible.

In this Indian camp also were found many relics of the Custer battle. So far as that is in question, I may say that I consider that action to have been a fair and square stand-up fight, in which one side was defeated and its members all died fighting.* Naturally, the Indians despoiled the slain for trophies. White soldiers have done the same when conditions have been reversed, as has been noted in the preceding chapters of this book. Of course, the Indians mutilated the dead and tortured the living, but some instances of both practices are found among white men, and we cannot judge the Indian by our standards, anyway.

But in the camp there were other evidences of savage ferocity, from which the soul shrinks in horror, and which showed that these Indians were among the most cruel and ruthless on the continent, and that they were only getting what they had given. Two instances will suffice. The troops took from the body of a dead warrior an unique necklace of human forefingers, which had been displayed with pride upon his barbaric breast;† and a bag was found which contained the right hands of twelve little Shoshone babies and children, which had been recently cut from little arms to give some ruthless warrior a ghastly trophy.

* See Preface for discussion of the term "Massacre."

† A picture of a similar necklace may be seen in Captain J. Lee Humfreville's interesting book, "Twenty Years Among Our Hostile Indians."

CHAPTER ELEVEN

Miles' Great Campaigning

I. Miles and His Foot Cavalry Defeat Sitting Bull

NOW let us turn to Miles and his men.

General Miles was ordered to march his command up the Yellowstone to the mouth of the Tongue River, and establish a temporary post or cantonment there for the winter. He was an officer in whom great confidence was reposed, and from whom much was to be expected. He had as brilliant a record in the Civil War as Custer, and had practically fought one decisive battle in the closing campaign on his own responsibility, with splendidly successful results. He was a natural-born soldier, and he never showed his talents to better advantage than in the operations which followed. His career before and after this period is still fresh in the minds of a grateful people.

While Crook and his men were hammering away in one portion of the field, Miles was doing splendid service in the other. The original intention had been to place under his command some fifteen hundred men, but the force he really received amounted only to about five hundred. With these he was not expected to do more than maintain his position, and acquire such information as he could in preparing for the spring and summer campaign of the following year. That was not, from his point of view, a satisfactory program.

Veteran Indian fighters in the Northwest informed him that it would be useless to try to reach the Indians in the winter; but Miles was not that kind of a soldier. If the Indians could live in tepees in that season, he saw no reason why white soldiers should not move against them in spite of the weather. He had one of the finest regiments of infantry in the service — the Fifth. Based upon the report of courts-martial, discipline, etc., no regiment surpassed or even equaled its record. Miles himself proved to be the most successful commander against Indians that the war produced, and his success was not due to what envious people called good luck. It was well merited and thoroughly earned.

The government, upon the representations of Sheridan and Sherman, which were based upon Miles' previous successful fighting with the Southwestern Indians, allowed the young colonel everything he asked for. If his troops were not completely equipped for the work in which their commander designed to employ them, it would be his fault. With wise forethought, he provided the soldiers as if for an arctic expedition. They cut up blankets for underwear. They were furnished with fur boots and the heaviest kind of leggings and overshoes. Every man had a buffalo overcoat and a woolen or fur mask to go over his face under his fur cap. Their hands were protected by fur gloves. It was well for them that they were thus provided, for the winter of 1876-7 was one of the most severe that had ever visited that section of the country. The mercury frequently froze in the thermometer, and on one occasion a temperature of sixty degrees below zero was recorded by the spirit thermometer.

Busying themselves during the late fall, which was, in effect, winter, in the erection of the cantonment on the Tongue and Yellowstone, the first important touch they got with the Indians was on the 18th of October, when Lieutenant-Colonel Elwell S. Otis, commanding a battalion of four companies of the Twenty-third Infantry, escorting a wagon-load of supplies from Glendive, Montana, to the cantonment, was attacked by a large force of hostiles. The attack was not delivered with any great degree of force at first, but it grew in power until the troops had to corral the train. The soldiers had a hard fight to keep the animals from being stampeded and the train captured. Having beaten off the Indians, the train advanced, fighting, until Clear Creek was reached. During a temporary cessation of the attacks a messenger rode out from the Indian lines, waving a paper, which he left upon a hill in line with the advance of the train. When it was picked up, Colonel Otis found it to be an imperious message — probably written by some half-breed — from the chief whom he had been fighting. It ran as follows:

"Yellowstone.

"I want to know what you are doing traveling on this road. You scare all the buffalo away. I want to hunt in this place. I want you to turn back from here. If you don't, I'll fight you again. I want you to leave what you have got here, and turn back from here.

I am your friend,

SITTING BULL.

"I mean all the rations you have got and some powder. Wish you would write me as soon as you can."

I consider this document unique in the history of Indian warfare, and it well illustrates not only the spirit, but the naïveté of the great chief. Otis despatched a scout to Sitting Bull with the information that he intended to take the train through to the cantonment in spite of all the Indians on earth, and if Sitting Bull wanted to have a fight, he (Otis) would be glad to accommodate him at any time and on any terms. The train thereupon moved out, and the Indians promptly recommenced the fight. But the engagement was soon terminated by a flag of truce. A messenger appeared, who stated that the Indians were tired and hungry and wanted to treat for peace. Otis asked Sitting Bull to come into his lines, but that wily old chief refused, although he sent three chiefs to represent him.

Otis had no authority to treat for peace or anything else, but he gave the Indians a small quantity of hardtack and a couple of sides of bacon, and advised them to go to the Tongue River and communicate with General Miles. The train then moved on, and after following a short distance, with threatening movements, the Indians withdrew.

On the same night Otis fell in with Miles and his whole force. Miles, being alarmed for Otis' safety, had marched out to meet him. The train was sent down to the cantonment, and the troops, numbering three hundred and ninety-eight, with one gun, started out in pursuit of Sitting Bull. They overtook him on the 21st of October at Cedar Creek. With Sitting Bull were Gall and other celebrated chiefs, and one thousand warriors of the Miniconjous, San Arcs, Brulés, and Unkpapas, together with their wives and children, in all over three thousand Indians. Crazy Horse, with the Oglalas and Two Moon's band of the Northern Cheyennes, were not with Sitting Bull, while Dull Knife's band, as we have seen, had gone to Wyoming for the winter.

The reason for this separation is obvious. They could better support the hardships of the winter, more easily find shelter, and with less difficulty escape from the pursuing soldiers, if they were broken up in smaller parties.

Sitting Bull asked Miles for an interview, which was arranged. He was attended by a subchief and six warriors, Miles by an aide and six troopers. The meeting took place between the lines, all parties being on horseback.

Sitting Bull wanted peace on the old basis. The Indians demanded permission to retain their arms, with liberty to hunt and roam at will over the plains and through the mountains, with no responsibility to any one, while the government required them to surrender their arms and come into the agencies. The demands were irreconcilable therefore. The interview was an interesting one, and although it began calmly enough, it grew exciting toward the end.

Sitting Bull, whom Miles describes as a fine, powerful, intelligent, determined looking man, was evidently full of bitter and persistent animosity toward the white race. He said no Indian that ever lived loved the white man, and that no white man that ever lived loved the Indian; that God Almighty had made him an Indian, but He didn't make him an Agency Indian, and he didn't intend to be one.

The manner of the famous chief had been cold, but dignified and courteous. As the conversation progressed, he became angry — so enraged, in fact, that in Miles' words "he finally gave an exhibition of wild frenzy. His whole manner seemed more like that of a wild beast than a human being. His face assumed a furious expression. His jaws were tightly closed, his lips were compressed, and you could see his eyes glisten with the fire of savage hatred."*

One cannot help admiring the picture presented by the splendid, if ferocious, savage. I have no doubt that General Miles himself admired him.

At the height of the conference a young warrior stole out from the Indian lines and slipped a carbine under Sitting Bull's blanket. He was followed by several other Indians to the number of a dozen, who joined the band, evidently meditating treachery. Miles, who, with his aide, was armed with revolver only, promptly required these new auxiliaries to retire, else the conference would be terminated immediately. His demand was reluctantly obeyed. After some further talk, a second meeting was appointed for the morrow, and the conference broke up.

During the night Miles moved his command in position to be able to intercept the movement of the Indians the next day. There was another interview with the picturesque and imperious savage, whose conditions of peace were found to be absolutely impossible, since they involved the abandonment of all the military posts, the withdrawal of all settlers, garrisons, etc., from the country. He wanted everything and would give nothing. He spoke like a conqueror, and he looked like one, although his subsequent actions were not in keeping with the part. Miles, seeing the futility of further discussion, peremptorily broke up the conference. He told Sitting Bull that he would take no advantage of the flag of truce, but that he would give him just fifteen minutes to get back to his people to prepare for fighting. Shouting defiance, the chiefs rode back to the Indian lines.

There was "mounting in hot haste" surely, and hur-

* Personal Recollections of General Nelson A. Miles, U. S. A.

ried preparations were made for immediate battle on both sides. Watch in hand, Miles checked off the minutes, and exactly at the time appointed he ordered an advance. The Indians set fire to the dry grass, which was not yet covered with snow, and the battle was joined amid clouds of flame and smoke. Although outnumbered nearly three to one, the attack of the soldiers was pressed home so relentlessly that the Indians were driven back from their camp, which fell into the possession of Miles.

The Sioux were not beaten, however, for the discomfited warriors rallied a force to protect their flying women and children, under the leadership of Gall and others, Sitting Bull not being as much of a fighter as a talker. They were led to the attack again and again by their intrepid chiefs. On one occasion, so impetuous was their gallantry that the troops were forced to form square to repel their wild charges. Before the battle was over — and it continued into the next day — the Indians had been driven headlong for over forty miles.

They had suffered a serious loss in warriors, but a greater in the destruction of their camp equipage, winter supplies, and other property. Two thousand of them came in on the third day and surrendered, under promises of good treatment. Several hundred broke into small parties and scattered. Miles' little force was too small to be divided to form a guard for the Indians who had been captured; and besides, he had other things to do, so he detained a number of the principal chiefs as hostages, and exacted promises from the rest that they would surrender at the Spotted Tail or Red Cloud Agency — a promise which, by the way, the great majority of them kept. Sitting Bull, Gall, and about four hundred others refused to surrender, and made for the boundary line, escaping pursuit for the time being.

This was the first and most serious defection from the Indian Confederacy. It was followed by others. In a subsequent campaign, in the depth of winter, a battalion under Lieutenant Baldwin struck Sitting Bull's depleted and starving camp on two separate occasions, inflicting further loss upon that implacable chieftain.*

II. Miles' Crushing Defeat of Crazy Horse at Wolf Mountain

Late in December Miles, having practically eliminated Sitting Bull from the game, moved out against Crazy Horse. He had with him five companies of the Fifth Infantry and two of the Twenty-second, in all four hundred and thirty-six officers and men and two Napoleon guns. These guns were fitted with canvas wagon-tops, and were so disguised as exactly to resemble the supply wagons of the train. The men left the cantonment on the 29th of December, 1876. It had been learned that Crazy Horse was in the valley of the Tongue River, south of the Yellowstone. There were sharp skirmishes on the first and third of January between the advance and war parties of Indians, who were moving gradually up the Tongue toward the mountains. On the evening of the 7th of January, 1877, a young warrior and a woman were captured, belonging to those Cheyennes who were still with Crazy Horse and the Unkpapas, and were

* As an instance of Miles' capacity in handling men, this is what Baldwin says in a private letter, afterward made public, of the orders he received: "When I was given command of this battalion opposite the mouth of Squaw Creek, and the General took command of a less number of men, it was a question as to which would find the hostile Indians, and with the only order or suggestion given by him in that earnest manner characteristic of him, he said, 'Now, Baldwin, do the best you can. I am responsible for disaster, success will be to your credit; you know what my plans are, and what we are here for.'" There is a dashing, manly ring about such words which I rejoice to recognize. It is a great soldier who can first choose and then trust his subordinates.

From the collection of J. Robert Coster

GEN. JOHN GIBBON — GEN. NELSON A. MILES

GEN. WESLEY MERRITT — GEN. ALFRED H. TERRY

SOME FAMOUS INDIAN FIGHTERS

related to some of the principal members of the band. From them much was learned of the situation of the Indian position.

The next morning, the weather being bitterly cold, the men moved out to attack the Indian camp. Crazy Horse's warriors numbered between eight and nine hundred. He had posted his men on the cliffs surmounting a valley in the Wolf Mountains, a spur of the Big Horn Range, not far from Crook's battle-ground on the Rosebud. The troops entered the valley in full view of the Indians occupying the heights. The position was well chosen; for in order to make the attack, the soldiers would have to climb straight up the walls to get at the Indians, who were enabled, by the configuration of the ground and by their numbers, almost to surround the soldiers. One reason why Crazy Horse was willing to fight was because of his great desire to get possession of the Indians recently captured.

Seeing that Crazy Horse was willing to accept battle, Miles made his preparations deliberately. The troops, out of range of the Indians, calmly had breakfast and made their camp secure. Having done everything at his leisure, Miles moved out to the attack.

The Sioux were plainly visible on the cliffs. They could be seen shaking their fists and brandishing their rifles as the soldiers slowly advanced through the deep snow which covered the ground. The Indians seemed absolutely confident that Miles was marching into a trap, that when he got into the cañon he would be unable to scale the slopes, and they would have him at their mercy. There was no ambush about it. The whole thing was open and plain. They had chosen their position and had invited the soldiers to make at them. There was, indeed, no other way for Miles to get to them, so cunningly had they taken advantage of the ground, except the way which lay open before them. As the troops drew nearer, the gestures of defiance and contempt were accompanied by yells and jeers. Among the things they shouted in their confident assurance of success were these significant words:

"You have had your last breakfast!"

Indeed, the grim prophecy did not seem unlikely of fulfilment.

It might have been supposed that men, encumbered as were the soldiers with their heavy, winter clothing, could never have scaled those heights, especially in the face of such opposition as the redoubtable warriors of Crazy Horse would offer. If they did not succeed in clearing the cliffs of the Indians, they would probably be shot down in scores in the valley. They would then be forced to retreat to their train, if any of them were left alive to do so, and stand a siege; and as they were three or four hundred miles from any possible relieving force, and in the depth of a Dakota winter, that would mean a speedy annihilation. It was a serious risk to take, but no battle was ever won without taking risks, and the nice art of the soldier consists in knowing what risks to take and when to take them. Not the least of Miles' claims to admiration as a commander was his determination, under all circumstances, to fight then and there.

Undaunted by the threatening prospect and unmoved by the savage shouts and jeers, although some of the scouts who knew the Sioux language retorted in kind, the troops deployed, and at as rapid a pace as they could manage, started for the hills. The artillery was exposed and unlimbered, and the shells thrown into the Indian position caused great surprise and consternation. The key to the position was a high elevation upon the left. The Indians who held it were led by Big Crow, the chief medicine man. As the battle began he exposed himself freely between the lines, dressed in a magnificent Indian war shirt and bonnet, running up and down and yelling like a fiend.

Miles massed a little column against Big Crow and the warriors defending the eminence. At the same time he ordered a general escalade of the cliff along the whole line. Under a heavy fire, which, however, like most plunging fires down the sides of mountains or slopes, did but little damage, the troops slowly toiled up the icy, snow-covered bluffs.* Led by Major Casey and Captains McDonald and Baldwin, the charge was delivered with the utmost resolution. It was not a dash. No men, encumbered as were those soldiers, could move rapidly up icy cliffs, covered, wherever the sharpness of the acclivity permitted, with from one to three feet of snow. It was rather a slow, dogged, determined crawl, with a stop every few moments to fire at some Indian silhouetted above them on the gray sky-line of that winter morning.

The fighting for the high cliff on the left of the line was spirited and desperate. Finally, the men came to a hand-to-hand struggle. The Indians clung tenaciously to the post until Big Crow was shot, when the soldiers succeeded in dislodging them. This bluff commanded the lines. It was occupied by the troops, who poured an enfilading fire upon the army of Crazy Horse. The Indian position, therefore, became untenable, and fighting

*At the battle of King's Mountain, in the American Revolution, the small loss of life among the Americans was due to the fact that the English, trained marksmen though they were, firing down the slopes of the mountain, overshot their opponents, although they had them in full view all the way up the slope; and it is the tendency of troops always to do the same thing. Troops on a level usually fire too low, and the ground between the advancing lines of soldiers is often plowed up by bullets from the depressed muzzles, which should have gone into the breasts of the approaching enemy.

sullenly and stubbornly, they withdrew in good order, though closely pursued by the troops. In the latter part of the advance snow began to fall, and before the battle was closed the combatants were fighting in the midst of a blinding storm. Miles says that the moment at which the Indians turned their backs and began the retreat was one in which he felt relief scarcely to be expressed, so desperate had been the fighting, so difficult the ascent, and so doubtful the result.

The Indians were pursued for some distance, and a large portion of their camp equipage, with supplies, was captured. On the whole, they had suffered a most disheartening and disorganizing defeat. Their ammunition was about gone, their confederates in other tribes had been captured, the main body of the redoubtable Cheyennes had been crushed and were starving, the Unkpapas, the Miniconjous, the Sans Arcs, and the Brulés had surrendered. The game was up. There was nothing for Crazy Horse and the exhausted remnant which remained faithful to him to do but to surrender, which they accordingly did in the early spring.

III. The Capture of Lame Deer's Village

There remained, then, in the field practically but one band of sixty lodges,* under Lame Deer and Iron Star, who refused positively to surrender. The indefatigable and brilliantly successful Miles pursued this band, overtook it, surprised it one morning in May, captured the village, dispersed the greater portion of the Indians, and succeeded in isolating and surrounding Lame Deer and Iron Star, with half a dozen principal warriors. Miles was very desirous of taking them alive. He advanced

* Each lodge accounted for from five to ten persons.

with some of his officers toward the desperate little body of Indians who had been cut off from the fleeing mass of savages, making peace signs and crying peace words.

The Indians were tremendously excited and remained on guard, but committed no act of hostility. Miles rode up, and leaning over the saddle, extended his hand to Lame Deer. The intrepid chieftain, who was quivering with emotion under his Indian stoicism, grasped the general's hand and clung to it tightly. Iron Star took Baldwin's hand. The other Indians came forward, reluctantly, with hands extended, and all was going well.

At this juncture one of the white scouts, not knowing what was going on, dashed up to the group, and possibly under a misapprehension that the life of the commanding officer was threatened, covered Lame Deer with his rifle. The Indian, probably thinking that he was to be killed in any event, resolved to die fighting. Miles strove to hold him and to reassure him, but by a powerful wrench he freed himself, lifting his rifle as he did so, and pointing it straight at the general.

Miles had been in many battles, but he was never nearer death than at that moment. His quickness and resource did not desert him. Just as the Indian's finger pressed the trigger he dug his spurs into his horse and swung the animal aside in a powerful swerve. Lame Deer's bullet, which missed him by a hair's breadth, struck one of the escort and instantly killed him. Iron Star also drew away from Baldwin and raised his rifle, as the other Indian had done. None of them were so quick, however, as Lame Deer had been. The soldiers closing in had seen Lame Deer's motion, and before any further damage was done by the Indians they were overwhelmed by a rapid fire, which stretched them all dead upon the ground. The fighting had been short, but exceedingly sharp. The troops lost four killed and seven wounded, the Sioux fourteen killed and a large number wounded. The band was completely broken up, and most of the Indians surrendered soon after.*

Of all the Indians who had borne prominent parts in this greatest of our Indian wars with the savage tribes, there remained at large only the indomitable Sitting Bull, and he had escaped capture because, with a wretched band of starving but resolute followers, he succeeded in crossing the British Columbia boundary line.

Crook's persistence, Mills' bold stroke, Mackenzie's desperate dash up Willow Creek Cañon, Miles' splendid campaigning, his hard fighting at Cedar Creek and Wolf Mountain, his pursuit of Lame Deer, his policy and skill in dealing with the critical situations which had arisen, at last brought peace to the blood-drenched land. The most important work ever done by the United States Army outside of the greater wars of the nation had been successfully and brilliantly accomplished.

IV. Farewell to a Great Chief and His Hopes

A note of the fate of the two chief antagonists of the United States may fittingly close this chapter. Sitting Bull returned to the United States, and surrendered to the army a few years later. Ever a malcontent, he was one of the moving spirits in the Ghost Dance uprising, which culminated in the battle of Wounded Knee in 1890, and he was killed by the Indian police while resisting arrest.†

* See close of this chapter for another account of the Lame Deer Fight.

† These affairs are to be discussed at length in a forthcoming volume.

The end of Crazy Horse came sooner, in a mêlée in a guard-house on the 7th of September, 1877. He was stabbed in the abdomen, and died from the effects of the wound. He was dissatisfied always, in spite of his surrender, and had been conspiring to take the war-path again. Believing that his intentions had become known and that he would be rigorously dealt with on account of the discovery, he started to run amuck, with a knife of which he had become possessed by some means, in the guard-house. When the fracas was over, he was found on the ground, with a desperate wound in the abdomen. Whether the wound was given by the bayonet of the sentry at the door, whether the blow was delivered by some of the Indians who threw themselves upon him, and with whom he struggled, is a matter which cannot be determined. However it was come by, it was enough, for from the effects he died in a short time.

So that was the melancholy end of Crazy Horse, the protagonist of these tales, and one of the most famous Indians that ever lived. Captain Bourke * thus describes him:

"I saw before me a man who looked quite young, not over thirty years old, five feet eight inches high, lithe and sinewy, with a scar in the face. The expression of his countenance was one of quiet dignity, but morose, dogged, tenacious, and melancholy. He behaved with stolidity, like a man who realized that he had to give in to Fate, but would do so as sullenly as possible. . . . All Indians gave him a high reputation for courage and generosity. In advancing upon an enemy, none of his warriors were allowed to pass him. He had made himself hundreds of friends by his charity toward the poor, as it was a point of honor with him never to keep anything

* "On the Border With Crook," Captain John G. Bourke, U. S. A.

for himself, excepting weapons of war. I never heard an Indian mention his name save in terms of respect. In the Custer Massacre, the attack by Reno had first caused a panic among the women and children and some of the warriors, who started to flee; but Crazy Horse, throwing away his rifle, brained one of the incoming soldiers with his stone war-club, and jumped upon his horse."

Crazy Horse was a born soldier, whose talents for warfare and leadership were of the highest order. He had repulsed Reynolds on the Powder River, wresting a victory from apparent defeat. He had thrown himself in succession upon the columns of Crook on the Rosebud and of Custer on the Little Big Horn ; and it must be admitted that he had not only checked, but had driven back, Crook by a crushing attack upon him, while he had annihilated half of Custer's command. He had fought a desperate, and, from a military point of view, highly creditable, action with Crook's vastly superior forces at Slim Buttes. The only man who had fairly and squarely defeated him was Miles at Wolf Mountain, and even there Crazy Horse managed to keep his force well in hand as he withdrew from the field.

He would probably never have surrendered, had it not been for the defections around him, and for the disastrous defeat of the Cheyennes by Mackenzie, and the destruction of so much of his camp equipage at Wolf Mountain. As it was, he might have continued the fighting, had not his warriors been freezing and starving, and almost entirely out of ammunition. There was nothing left for the Indians but surrender. As one of them said to Miles:

"We are poor compared with you and your force. We cannot make a rifle, a round of ammunition, or a knife. In fact, we are at the mercy of those who are taking possession of our country. Your terms are harsh and cruel, but we are going to accept them, and place ourselves at your mercy."

That summed up the situation, although the terms granted the Indians were very far from being harsh or cruel.

So passed out of history the great war chief of the Sioux, one of the bravest of the brave, and one of the most capable and sagacious of captains in spite of his absurd name. He had many of the vices, perhaps all the vices, of his race; but he had all their rude virtues, too, and great abilities, which most of them lacked. Sitting Bull, wise, crafty, indomitable as he was, was not to be compared with him for a moment.

It was a tragedy any way you look at it. You cannot but feel much admiration for those Sioux and Cheyennes — cruel, ruthless though they were. I bid good-by to them with a certain regret.

Some one has said, as the Rosebud and the Little Big Horn marked the high-water of Indian supremacy in the Northwest, so the forgotten grave of Crazy Horse marks an ebb from which no tide has ever risen.

As he passes to the happy hunting-ground in the land of the Great Spirit, I stand and salute him with a feeling of respect which I have gathered not only from a study of his career, but from the statements and writings of men who could best judge of his qualities—for they were the soldiers who fought him.

NOTES ON THE LAME DEER FIGHT

By Colonel D. L. Brainard, U. S. A.*

The command, consisting of four troops of the Second Cavalry, "F," "G," "H," and "L," two companies of the Fifth Infantry, two

* Colonel Brainard won his commission by his heroic conduct in the Greely Arctic Expedition, 1881–4. — C. T. B.

of the Twenty-second Infantry, and a company of mounted scouts, all under command of Colonel Nelson A. Miles, left the cantonment on Tongue River May 1, 1877, and marched up Tongue River, with a view of intercepting a band of hostile Indians, under Lame Deer, known to be at or near the head-waters of the Rosebud River. The transportation consisted of bull teams, mule teams, and a few pack animals. The command marched up Tongue River four days, when the train was left in charge of a small guard, the main command pushing on with pack trains, the cavalry leading and the infantry following more slowly, striking across country toward the Rosebud River, marching day and night, stopping only long enough to make coffee for the men, and to rest and graze the animals.

We bivouacked on the evening of the 6th in a deep valley near Little Muddy Creek, and about two o'clock the following morning were again in the saddle, moving silently and swiftly down the valley toward the Indian camp, which had been located the previous evening by White Bull, Brave Wolf, Bob Jackson, and the other scouts. The scouts had reported that the camp was only about six miles distant, but it was soon discovered that it was much farther than this, and at early dawn we were still some distance away. The command had been moving at a trot, but the gallop was immediately taken up, and just as the sun appeared above the horizon, we rounded a bend in the valley and came in sight of the Indian camp, which was located on the right side, close to the hills.

At first we saw no Indians except a few boys guarding the ponies, which were grazing a little distance beyond the camp, but they came out immediately, and dropping in the grass, began to fire in our direction, though without effect. As we charged down on the camp, these Indians, together with squaws and children, ran for the hills, driving with them the few horses that were near the tepees. "H" Company, under command of Lieutenant Lovell H. Jerome, charged through the camp and beyond, capturing the pony herd. The other companies, all under command of Captain Ball, charged to the village, formed line to the right, deployed as skirmishers, and pursued the Indians up the hill.

The hills were so steep at this point that it was necessary to dismount the command and advance on foot, the horses being sent around by an easier route to join us later near the summit of the hill. The line as formed was "F" troop (Tyler) on the right, "L" troop (Norwood) center, and "G" troop (Wheelan) on the left. The Indians were driven up over the hills, where they scattered like quail. Our horses were brought up, and mounting, we charged across the country for two or three miles, and later returned to the village.

As my recollections serve me, four soldiers and fourteen Indians were killed, ten soldiers being wounded, myself being one of the number. About four hundred ponies were captured, which were afterward used for mounting a batallion of Infantry, which later performed much effective work in the field. There were over sixty tepees, in which we found tons of dried buffalo meat, a few arms, some ammunition, and a great many buffalo robes, saddles, and an assortment of camp property, all of which were burned that afternoon, thus so effectually crippling the band that the remnant came in and surrendered a few weeks later.

We camped on the battle-ground that night, the following day moving back in the direction of our wagon train.

One of the most interesting incidents of the fight occurred just as the troop to which I belonged ("L") charged on the village. I saw General Miles riding toward the first tepee, near which were two Indians, followed by his orderly. He called out something to these Indians which I did not understand, but I later understood he had called on them to surrender. One of these was evidently the Chief, Lame Deer, for he wore a long head-dress of eagle feathers, the head-dress reaching to the ground. As Miles approached on horseback, the Chief walked rapidly toward him, with his hand extended, as though to shake hands, but when within ten or twelve feet of him, the Indian in the rear, who was said to be Iron Star, a son of Lame Deer, and also a medicine man of the tribe, called sharply to Lame Deer, presumably warning him of the approaching troops, and urging him to follow the other Indians to the hills.

Lame Deer stopped, turned, hesitated, then ran back a few steps, and picking up a loaded carbine from the ground, fired point blank at General Miles, who, seeing the movement, wheeled his horse sharply and bent forward. The bullet passed over him, striking his orderly in the breast, killing him instantly. The Chief then ran up the steep hill, accompanied by the other Indian. The head-dress made a very conspicuous mark, and many shots were immediately fired in that direction. From his tottering steps we saw that the Chief was badly wounded, and at this point his companion, instead of escaping as he could have done, placed his arm around the Chief's waist, and supported him up the hill. About this time the Chief drew a revolver, and without turning about, held it in rear of him and fired in our direction, the bullets striking the ground only a few feet in his rear. This act, we assumed, was one of defiance of a man who knew he could not escape, but who was game to the last. Iron Star supported the Chief until the latter fell, when he escaped over the hill, only to be killed by "G" troop, which had been pushing up on that side. After the devotion and bravery he had displayed in supporting Lame Deer up the hill, we were almost sorry he had not escaped alive.

A few days later Bob Jackson told me that, on examining Lame Deer's body after the fight, he had found that he had been hit seventeen times.

Another incident which illustrates the valor of the United States soldier was that of Private Leonard, Troop "L," Second Cavalry, who had dropped behind to readjust his saddle, a couple of miles from the Indian camp. The command was moving rapidly, and the Indians slipped in between the rear of the column and this lone soldier. However, when he saw them he rode to the top of a hill, and lying down behind some rocks, held these Indians at bay for several hours until relief came to him. It was fortunate that relief came as it did, for he had nearly exhausted his ammunition in firing at these Indians, who had several times charged his position.

CHAPTER TWELVE

What They Are There For
A Sketch of General Guy V. Henry, a Typical American Soldier

I. Savage Warfare

THE most thankless task that can be undertaken by a nation is warfare against savage or semi-civilized peoples. In it there is usually little glory; nor is there any reward, save the consciousness of disagreeable duty well performed. The risk to the soldier is greater than in ordinary war, since the savages usually torture the wounded and the captured. Success can only be achieved by an arduous, persistent, wearing down process, which affords little opportunity for scientific fighting, yet which demands military talents of the highest order.

Almost anybody can understand the strategy or the tactics of a pitched battle where the number engaged is large, the casualties heavy, and the results decisive; but very few non-professional critics appreciate a campaign of relentless pursuit by a small army of a smaller body of mobile hostiles, here and there capturing a little band, now and then killing or disabling a few, until in the final round-up the enemy, reduced to perhaps less than a score, surrenders. There is nothing spectacular about the performance, and everybody wonders why it took so long.

And as injustice and wrong have not been infrequent in the preliminary dealings between the government and the savages, the soldier, who has only to obey his orders, comes in for much unmerited censure from those who think darkly though they speak bitterly. Especially is he criticized if, when maddened by the suffering, the torture, of some comrade, the soldier sinks to the savage level in his treatment of his ruthless foeman. No one justifies such a lapse, of course, but few there be who even try to understand it. The incessant campaigning in the Philippines, with its resulting scandals, is an instance in point.

Long before the Spanish-American War and its Philippine corollary, however, our little army had shown itself capable of the hardest and most desperate campaigning against the Indians of the West — as difficult and dangerous a work as any army ever undertook. There was so much of it, and it abounded with so many thrilling incidents, that volumes could be written upon it without exhausting its tragedy, its romance. There were few soldiers who served beyond the Mississippi from 1865 to 1890 who did not participate in a score of engagements, whose lives were not in peril more than once in many a hard, but now forgotten, campaign.

One of the bravest of our Indian fighters was Guy V. Henry. Personally, he was a typical representative of the knightly American soldier. Officially, it was his fortune to perform conspicuous services in at least three expeditions subsequent to the Civil War. He was a West Pointer, and the son of another, born in the service

From the collection of J. Robert Coster

COL. RANALD S. MACKENZIE — GEN. GUY V. HENRY

CAPT. ANSON MILLS — W. F. CODY (BUFFALO BILL)

GROUP OF DISTINGUISHED INDIAN FIGHTERS

All except General Henry from contemporary photographs

at Fort Smith in the Indian Territory. Graduating in 1861, a mere boy, he participated in four years of the hardest fighting in the Civil War, from Bull Run to Cold Harbor. At the age of twenty-three, his merit won him the appointment of Colonel of the Fortieth Massachusetts Volunteers, "a regiment that was never whipped." The tall, brawny Yankees fairly laughed at the beardless stripling who was appointed to command them. "They laugh best who laugh last," and Henry had the last laugh. He mastered them, and to this day they love his memory.

He was thrice mentioned in despatches, and brevetted five times for conspicuous gallantry in action during the war, out of which he came with the rank of brigadier-general. For heroic and successful fighting at Old Cold Harbor, he received the highest distinction that can come to a soldier, the medal of honor. Having two horses shot from under him in the attack upon the Confederate lines, he seized a third from a trooper, mounted him under a withering fire, and led his soldiers forward in a final assault, which captured the enemy's intrenchments; this third horse was shot under him just as he leaped the breastworks.

"Thin as a shoestring and as brave as a lion," he was a past-master of military tactics and a severe disciplinarian. "I tell you he is a martinet," cried one young officer angrily, smarting under a well-deserved reproof. "You are wrong," replied a wiser officer, who knew Henry better; "he is trying to make your own record better than you could ever make it yourself." Sudden as a thunderbolt and swift as a hawk when he struck the red Sioux, in his family and social relations he was a kindly, considerate, Christian gentleman. He could kill Indians — but never cruelly, mercilessly, only in open warfare — and teach a class in Sunday-school. I've seen him do the latter, and no man did it better; the boys of his class simply idolized him. And his men in the army did the same. Cool and tactful, a statesman, for all his fiery energy, he was perhaps the best of our colonial governors. When he died the people of Porto Rico mourned him as a friend, where the little children had loved him as a father.

II. A March in a Blizzard

At the close of the Civil War he was transferred to the Third Cavalry, a regiment with which he was destined to win lasting renown. It must have been hard for men who had exercised high command, and who had proved their fitness for it, to come down from general officers to subalterns; but Henry accepted the situation cheerfully. He was as proud of his troop of cavalry as he had been of his regiment and brigade of volunteers. His new detail took him to Arizona, where for two years he commanded a battalion engaged in hard scouting among the Apaches. The winter of 1874 found him at Fort Robinson in the Black Hills. While there he was ordered to go into the Bad Lands to remove certain miners who were supposed to be there in defiance of treaty stipulations.

The day after Christmas, with his own troop and fifteen men of the Ninth Infantry under Lieutenant Carpenter, with wagons, rations, and forage for thirty days, the men set forth. The expedition involved a march of three hundred miles, over the worst marching country on the face of the globe, and in weather of unimaginable severity, the cold continually ranging from twenty to forty degrees below zero. The miners were not found, and on the return journey the command, which had suffered terrible hardships, was overtaken by a blizzard.

When, in an Eastern city, the thermometer gets down to the zero mark, and it blows hard, with a heavy snow for twenty-four hours, people who are not familiar with the real article call such insignificant weather manifestations a blizzard. Imagine a fierce gale sweeping down from the north, filled with icy needles which draw blood ere they freeze the naked skin, the thermometer forty degrees below zero, a rolling, treeless country without shelter of any sort from the blinding snow and the biting wind, and you have the situation in which that expedition found itself. The storm came up an hour after breaking camp on what was hoped to be the last day on the return journey. To return to the place of the camp was impossible. To keep moving was the only thing to be done. The cold was so intense that it was at first deemed safer to walk than ride. The troops dismounted and struggled on. Many of the men gave out and sank exhausted, but were lifted to their saddles and strapped there, Henry himself doing this with his own hands. Finally, the whole party got so weak that it was impossible for them to proceed. In desperation, they mounted the exhausted horses and urged them forward. Henry had no knowledge of direction, but trusted to the instincts of his horse. He led the way. Many of the men had to be beaten to keep them awake and alive — to sleep was death.

Finally, when hope and everything else was abandoned, they came to a solitary ranch under the curve of a hill, occupied by a white man and his Indian wife. They were saved; that is, they had escaped with their lives. The horses were put in shelter in the corral, the men crowded into the house, and the painful process of thawing out was begun. The ranch was fifteen miles from Fort Robinson, and when the blizzard abated the next day wagons and ambulances were sent out, and the helpless soldiers were carried back to the post.

Most of them were in a terrible condition, and few had escaped. They were broken from the hardships they had undergone, especially from the freezing, which those who have suffered from declare causes a prostration from which it is difficult to recover. When Henry entered his quarters his wife did not recognize him. His face was black and swollen. His men cut the bridle reins to free his hands, and then slit his gloves into strips, each strip bringing a piece of flesh as it was pulled off. All his fingers were frozen to the second joint; the flesh sloughed off, exposing the bones. One finger had to be amputated, and to the day of his death his left hand was so stiff that he was unable to close his fingers again. As he was a thin, spare man, with no superfluous flesh, he had suffered more than the rest. Yet he made no complaint, and it was only due to his indomitable persistence that the men were not frozen to death that awful day. Henry's winter march is still remembered, by those in the old service, as one of the heroic achievements of the period.

III. A Ghastly Experience

The heroism and sufferings of the young soldier were nothing, however, to what he manifested and underwent two years later. Just before the Custer Massacre, General Crook, with some eleven hundred men, moved out from Fort Fetterman, Wyoming, on the expedition that culminated in the battle of the Rosebud. Colonel W. B. Royall had command of the cavalry of Crook's little army. One morning in June the Sioux and the Cheyennes, under Crazy Horse, who as a fighter and general was probably second to few Indians that ever lived, attacked Crook's men. The left wing, under Royall, was isolated in a ravine and practically surrounded by a foe who outnumbered them five to one. The rest of the army, heavily engaged, could give them no succor. The Indians made charge after charge upon the troops, who had all dismounted except the field officers. Henry had command of the left battalion of Royall's force. Cool as an iceberg, he rode up and down the thin line, steadying and holding his men. At one time, by a daring charge, he rescued an imperiled company under a brother officer.

At last, in one of the furious attacks of the Sioux, he was shot in the face. A rifle bullet struck him under the left eye, passed through the upper part of his mouth under the nose, and came out below the right eye. The shock was terrific. His face was instantly covered with blood, his mouth filled with it. He remained in the saddle, however, and strove to urge the troops on. In the very act of spurring his horse forward to lead a charge, he lost consciousness, and fell to the ground.

At that instant the war-bonneted Indians, superbly mounted, delivered an overwhelming onslaught on the left flank of the line. The men, deprived of their leader, for a time gave back. The Indians actually galloped over the prostrate figure of the brave soldier. Fortunately, he was not struck by the hoofs of any of the horses. A determined stand by Chief Washakie, of the friendly Shoshones, our Indian allies in that battle, who with two or three of his braves fought desperately over Henry's body, prevented him from being scalped and killed.

The officers of the Third speedily rallied their men, drove back the Indians, and reoccupied the ground where Henry lay. He was assisted to his horse and taken to the rear where the surgeons were. Such was the nature of his wound that he could not speak above a whisper; he could not see at all, he could scarcely hear, and he had great difficulty in breathing. As the doctor bent over him he heard the wounded man mumble out, "Fix me up so that I can go back!"

There was no going back for him that day. Through the long day he lay on the ground while the battle raged about him. There was little water and no shelter; there wasn't a tent in the army. Although it was bitter cold during the nights in that country at that season, at midday it was fearfully hot. He was consumed with thirst. His orderly managed to give him a little shade by holding his horse so that the shadow of the animal's body fell upon the wounded man. His wound was dressed temporarily as well as possible, and then he was practically left to die.

One of the colonel's comrades came back to him during a lull in the fight. There he lay helpless on the bare ground, in the shadow of the restive horse, which the orderly had all he could do to manage. No one else could be spared from the battle line to attend to Henry's wants, although, as a matter of fact, he expressed no wants. The flies had settled thickly upon his bandaged face. The officer bent over him with an expression of commiseration.

"It's all right, Jack," gurgled out from the bleeding lips; "it's what we're here for."*

* John F. Finerty, who was present as the correspondent of the Chicago *Times*, and who relates the incident, says that Henry, immediately after this remark, advised Finerty to join the army. Encouraging circumstances to back up such a recommendation!

Royall's forces were finally able to effect a junction with the main body by withdrawing fighting, and Henry was carried along any way in the hurried movement. The Indians at last withdrew from the field (the battle must be considered a drawn one), and then there was time to consider what was to be done with the wounded. The facilities for treatment were the slenderest. The column had been stripped of its baggage, in order to increase its mobility, to enable it to cope with the Indians. All they had they carried on their persons, and that included little but the barest necessities.

Nobody expected Henry to survive the night. He didn't expect to live himself, as he lay there through the long hours, listening to the men digging graves for those who had fallen, and wondering whether or not he was to be one of the occupants thereof. The next day they sent him to the rear. He was transported in what is called a travois. Two saplings were cut from the river bank; two army mules, one at each end, were placed between the saplings, which were slung over the backs of the animals. An army blanket, or piece of canvas, was then lashed to the poles, and on them the sufferer was placed. There were a number of wounded — none of them, however, so seriously as Henry. It was some two hundred miles to Fort Fetterman, and they carried him all that distance that way.

The weather at night was bitterly cold. In the daytime it was burning hot. The travois was so short — they had to take what poles they could get, of course — that several times the head of the rear mule hit the wounded officer's head, so that finally they turned him about, putting his head behind the heels of the foremost animal, where he was liable to be kicked to death at any moment.

On one occasion one of the mules stumbled and fell and pitched Henry out upon his head. The officers of the little escort stood aghast as they saw him fall out; but it is a matter of record, solemnly attested, that such was Henry's iron self-control that he made no sound, although the agony was excruciating. In fact, on the whole journey he made no complaint of any sort. His only food was broth, which was made from birds shot by the soldiers as they came upon them, and he got this very infrequently.

Finally, the little cortège reached Fort Fetterman. The last mishap awaited them there. The river was crossed by a ferryboat, which was pulled from shore to shore by ropes and tackles. The river was very high and the current running swiftly, and as they prepared to take the wounded officer across, the ropes broke, and the whole thing went to pieces, leaving him within sight, but not within reach, of clean beds, comforts, and medical attention he hoped to secure. Some of the escort, rough soldiers though they were, broke into tears as they saw the predicament of their beloved officer. He himself, however, true to his colors, said nothing. Finally, they offered to take him across the raging torrent in a small skiff — the only boat available — if he were willing to take the risk. Of course, if the skiff were overturned, he would have been drowned. He took the risk, and with two men to paddle and an officer to hold him in his arms, the passage was made.

IV. An Army Wife

Three hundred miles away, at Fort D. A. Russell, his wife was waiting for him. Long before he reached

Fort Fetterman, she heard through couriers the news of his wound, which was reported to her as fatal, although he had taken care to cause a reassuring message to be sent her with the first messenger. With the heroism of the army wife, although she was in delicate health at the time, she immediately made preparations to join him. The railroad at that time ran as far as Medicine Bow. Beyond that there was a hundred-mile ride to Fort Fetterman. All the troops were in the field; none could be spared from the nominal garrisons for an escort. Again and again Mrs. Henry made preparations to go forward, several times actually starting, and again and again she was forbidden to do so by the officers in command at the various posts. It was not safe to send a woman across the country with a few soldiers; the Indians were up and out in all directions. There was no safety anywhere outside the forts or larger towns; she had to stay at home and wait. Sometimes the devoted wife got word from her husband, sometimes she did not. The savages were constantly cutting the wires. Her suspense was agonizing.

Finally, the arrival of troops at Fort Fetterman enabled a stronger escort to be made up, and Henry was sent down to Fort D. A. Russell. The troops arrived at Medicine Bow on the third of July. The train did not leave until the next day. They were forced to go into camp. The cowboys and citizens celebrated the Fourth in the usual manner. That night the pain-racked man narrowly escaped being killed by the reckless shooting of the celebrators. Two bullets passed through the tent in which he lay, just above his head. The next morning found him on the train. His heart action had been so weakened by chloral and other medicines which they had given him, that at Sherman, the highest point on the journey, he came within an inch of dying.

His thoughts all along had been of his wife. When he got to the station he refused to get in an ambulance, in order to spare her the sight of his being brought home in that way. A carriage was procured, and supported in the arms of the physician and his comrades, he was driven back to the fort. With superhuman resolution, in order to convince his wife that he was not seriously hurt, he determined to walk from the carriage to the door. Mrs. Henry had received instructions from the doctor to control herself, and stood waiting quietly in the entrance.

"Well," whispered the shattered man, as she took him tenderly by the hand, alluding to the fact that it was the Fourth of July, "this is a fine way to celebrate, isn't it?"

After the quietest of greetings — think of that woman, what her feelings must have been! — he was taken into the house and laid on the sofa. The doctor had said that he might have one look at his wife. The bandages were lifted carefully from his face, so that he might have that single glance; then they were replaced, and the wife, unable to bear it longer, fled from the room. The chaplain's wife was waiting for her outside the door, and when she got into the shelter of that good woman's arms she gave way and broke down completely.

"You know," said the chaplain's wife, alluding to many conversations which they had had, "that you asked of God only that He should bring him back to you, and God has heard that prayer."

Everybody expected that Henry would die, but die he did not; perhaps it would be better to say die he would not. And he had no physique to back his efforts, only an indomitable will. He never completely recovered from that experience. He lost the sight of one eye permanently, and to the day of his death was liable to a hemorrhage at any moment, in which there was grave danger of his bleeding to death.

He took a year's leave of absence, and then came back to duty. In 1877, when his troop was ordered to the front in another campaign under Crook against the redoubtable Sioux, he insisted upon accompanying them. He had been out an hour or so when he fell fainting from the saddle. Did they bring him back? Oh, no! He bade them lay him under a tree, leave two or three men with him to look after him, and go ahead. He would rejoin them that night, when it became cooler and he could travel with more ease! What he said he would do he did. A trooper rode back to the post on his own account and told of his condition. An order was sent him by the post commander to return. Henry quietly said he would obey the first order and go on. He remained with the troop for six weeks, until finally he was picked up bodily and carried home, vainly protesting, the doctor refusing to answer either for his eyesight or for his life if he stayed in the field any longer.

V. The Buffalos and Their Famous Ride

Thirteen years after that Henry was commanding the Ninth Cavalry, with headquarters at Fort McKinney. The Ninth Cavalry was a regiment of negroes. From the overcoats which they wore in Wyoming in the winter they were called the "Buffalos," and sometimes they were facetiously referred to as "Henry's Brunettes."

Whatever they were called, they were a regiment of which to be proud.

In 1890 occurred the last outbreak of the Sioux under the inspiration of the Ghost Dancers, which culminated in the battle of Wounded Knee. Troops from all over the United States were hurried to the Pine Ridge Agency as the trouble began. On the 24th of December Henry and the "Brunettes" were ordered out to the former's old stamping ground in the Black Hills on a scouting expedition. It was bitter cold that Christmas Eve, but, thank God! there was no blizzard. Fifty miles on the back of a trotting horse was the dose before them. They rested at four A.M. on the morning of Christmas Day. Some of the garments the men wore were frozen stiff. They had broken through the ice of the White River in crossing it. How the men felt inside the frozen clothing may be imagined. Eight miles farther they made their camp. They did not have much of a Christmas celebration, for as soon as possible after establishing their base at Harney Springs they went on the scout. They hunted assiduously for several days, but found no Indians. These had gone south to join their brethren concentrated about the agency. One day they rode forty-two miles in a vain search. They got back to camp about seven o'clock. At nine a courier from the agency fifty miles away informed them of the battle of Wounded Knee, and that five thousand Oglala Sioux were mustering to attack the agency.

"Boots and Saddles!" instantly rang out, and the tired troopers mounted their jaded horses again. This time the camp was broken for keeps, and tents were struck, wagons packed to abandon it. It was a bitter cold night. There was a fierce gale sweeping through the valley, blowing a light snow in the faces of the men wrapped to the eyes in their buffalo coats and fur caps. They pushed steadily on in spite of it, for it was Henry's intention to reach the agency in the dark in order to avoid attack by the Indians.

It was thought advisable, therefore, to leave the wagon train under an escort of one company and press forward with the rest. The men arrived at the agency at daybreak, completing a ride of over ninety miles in less than twenty hours. Fires were kindled, horses picketed, and the exhausted men literally threw themselves on the ground for rest. They had been there but a short time when one of the men from the escort came galloping madly in with the news that the wagon-train was heavily attacked, and that succor must be sent at once. Without waiting for orders, without even stopping to saddle the horses, Henry and his men galloped back over the road two miles away, where the escort was gallantly covering the train. A short, sharp skirmish, in which one man was killed and several wounded, drove back the Indians, and the regiment brought in the train.

It was ten o'clock now, and as the negro troops came into the agency, word was brought that the Drexel Mission, seven miles up the valley, was being attacked, and help must be sent immediately. There were two regiments of cavalry available, the Seventh and the Ninth. For some unexplained reason, the Ninth was ordered out. In behalf of his men, Henry made protest. They must have a little rest, and so the Seventh was despatched, and was soon hotly engaged. Two hours later a messenger reported that the Seventh, in the valley where the mission was situated, was heavily attacked by the Indians, who had secured commanding positions on the surrounding ridges. Unless they could be relieved, they would probably be overwhelmed. Again the trumpet call rang out, and the tired black troopers once more climbed into their saddles and struck spurs into their more tired horses, galloping away to the rescue of their hard-pushed white comrades. The ridges were carried in most gallant style, and after some sharp fighting the Indians were driven back. The Seventh was extricated and the day was saved.

In thirty-four hours of elapsed time, the Ninth Cavalry had ridden one hundred and eight miles — the actual time in the saddle being twenty-two hours. They had fought two engagements and had rested only two hours. Marvelous to relate, there wasn't a sore-backed horse in the whole regiment. One horse died under the pressure, but aside from that and their fatigue, horses and men were in excellent condition.

That was probably the most famous ride ever performed by troops in the United States. For it Henry was recommended for a further brevet, as major-general — the sixth he had received.

The Spanish-American War was too short to afford Henry an opportunity to distinguish himself in the field, but in Porto Rico he showed that his talents were not merely of the military order. In the brief period his health permitted him to remain there, he accomplished wonders, and did it all in such a way as to gain the respect — nay, the affection — of the people over whom, with single-hearted devotion and signal capacity, he ruled. He stayed there until he broke down. I, sick with typhoid fever on a transport at Ponce, saw him just before he collapsed. We were old friends, and he came off to the ship to visit me. I was not too ill then to realize that his own time was coming. He would not ask to be relieved.

"Here I was sent," he said; "here I will stay until my duty is done."

He was the knightliest soldier I ever met, and I have met many. He was one of the humblest Christians I ever knew, and I have known not a few. It was his experience at Porto Rico which finally brought about his death; for it is literally true that he died, as a soldier should, in his harness. In those trying times at Ponce, when life and health were at a low ebb, he wrote, in the sacred confidence of his last letter to his faithful wife, words which it was not his custom to speak, but which those of us who knew him felt expressed his constant thought:

"I am here alone. One by one my staff officers have fallen ill and gone home. Home! — let us not speak of it. Jesus is here with me, and makes even this desolation home until a brighter one is possible."

So, his memory enshrined in the hearts that loved him, his heroic deeds the inspiration of his fellow-soldiers, passed to his brighter home Guy V. Henry, a Captain of the Strong

APPENDICES

APPENDIX A.

Being a Further Discussion of General Custer's Course in the Little Big Horn Campaign.*

I.

WHETHER General Custer did, or did not, obey General Terry's orders; whether these orders were, or were not, well considered, and such as could be carried out; whether, if General Custer did disobey General Terry's orders he was warranted in so doing by the circumstances in which he found himself, are questions of the deepest interest to the student of military matters and the historian thereof. I presume the problem they present will never be authoritatively settled, and that men will continue to differ upon these questions until the end of time.

The matter has been discussed, pro and con, at great length on many occasions. A number of books and magazine articles have been written upon different phases of the situation. I have come to the conclusion indicated in my own article, as I said, against my wish. In view of his heroic death in the high places of the field, I would fain hold General Custer, for whom I have long cherished an admiration which I still retain, entirely innocent. I have only come to this conclusion after a rigid investigation including the careful weighing of such evidence as I could secure upon every point in question.

This evidence consists, first, of a great variety of printed matter; second, of personal conversations with soldiers and military critics, which, as any record of it would necessarily be hearsay and second-hand, I have not set down hereafter save in one instance; third, of let-

*** All notes in this appendix are signed by the initials of their writers to identify them.—C. T. B.**

ters which have been written me by officers who, from their participation in the campaign, or from unusual opportunities to acquire knowledge concerning it which they have enjoyed, have become possessed of information which they were willing to give to me.

The object of this appendix is to set down, so that it may be here preserved in permanent and available form for future reference, such evidence in these letters as may be pertinent and useful; also to refer the student, who desires to go deeper into the subject, to some of the more valuable printed accounts which are easily accessible.

I am glad that some of the communications I have received, notably those from Colonel Godfrey, make a stout defense of General Custer. Perhaps upon consideration of Colonel Godfrey's points and arguments, which are not only strong and well taken, but also admirably put, the critic may be inclined to differ from my conclusion. For the sake of General Custer's fame, I sincerely hope so. I should be glad to be proved to be mistaken.

Without specifically noting the various descriptions of the campaign and battle, which are interesting, but irrelevant to my purpose,* Custer's conduct has been critically considered at some length—by persons whose standing requires that their opinions should be respectfully received — in several publications which I note in such order as best serves the purpose of this discussion without regard to the order in which they appeared.

Colonel Edward S. Godfrey,† U. S. A., now commanding the Ninth Cavalry, who, as a lieutenant, commanded K Troop, in Benteen's battalion, which joined that of Reno in the battle of the 26th of June, 1876, wrote a most interesting account of the battle, containing some valuable reflections upon some disputed points, which was published in the *Century Magazine*, Vol. XLIII., No. 3, January, 1892. To this article, in the same number, were appended certain comments by Major-General James B. Fry, U. S. A., since deceased.

This article and these comments came to the notice of Major-General Robert P. Hughes, U. S. A. (retired), then Colonel and Inspector-General. General Hughes was General Terry's aide-de-camp during the Little Big Horn Campaign. He wrote an exhaustive criticism on

* Such as Congressman Finerty's graphic account in his book, "War-path and Bivouac;" Dr. Charles S. Eastman's paper in the *Chautauqua Magazine*, Vol. XXXI., No. 4, 1900; and Mr. Hamlin Garland's report of Two Moon's account of the battle in *McClure's Magazine*, Vol. XI., No. 5, September, 1898. — C. T. B.

† General G. A. Forsyth writes me that he considers Colonel Godfrey one of the ablest officers in the United States Army — in which opinion I concur. — C. T. B.

Fry's comments to Godfrey's article, which was in effect a discussion of the main proposition that Custer disobeyed his orders and thereby precipitated the disaster, for which he was therefore responsible. This campaign was also considered in an article by Dr. E. Benjamin Andrews, president of the University of Nebraska, who was then president of Brown University, Providence, Rhode Island, which appeared in *Scribner's Magazine* for June, 1895. A fuller reference to Dr. Andrews' position will be made later.

General Hughes' article was offered to the *Century*, but was not accepted, and was finally published in the *Journal of the Military Service Institution*, Vol. XVIII., No. 79, January, 1896.

Among the many books in which the matter has been discussed, three only call for attention.

In "THE STORY OF THE SOLDIER," by Brigadier-General George A. Forsyth, U. S. A. (retired), the following comment appears:

"Under the peculiar condition of affairs, bearing in mind the only information he could possibly have had concerning Sitting Bull's forces, was Custer justified, in a military sense and within the scope of his orders, in making the attack?

"In the opinion of the writer he was within his orders, and fully justified from a military standpoint in so doing."

General Forsyth gives no reason for his decision, but it is to be presumed that he did not arrive at that decision hastily and carelessly, and as he is a very able and distinguished officer and military critic, due weight should be accorded his views.

In "THE UNITED STATES IN OUR OWN TIME," by Dr. E. Benjamin Andrews, published by Chas. Scribner's Sons, edition of 1903, pages 190-1-2-3, there is a concise discussion of the question, based on the article in *Scribner's Magazine*, referred to above, with some additional reflections on General Hughes' paper.

In "PERSONAL RECOLLECTIONS OF GENERAL NELSON A. MILES, U. S. A.," chapter xv., pages 198-210, there is a further discussion by the Lieutenant-General, lately in command of the United States Army.

In order clearly to understand what follows the student should refer to each of the sources mentioned and examine carefully into what is therein set forth. It is not practicable to quote all these authors at length. I have corresponded with every one of the authors mentioned except General Fry. I print their letters to me, having made no change except once in a while breaking a page into paragraphs and supplying a missing word here and there which had no especial bearing upon the point at issue. Some of the letters were written in pencil amid press of duties. Most of these documents I print without comment. It is necessary, however, that I should call attention to some features brought out by the correspondence.

President Andrews says, in the book referred to:

"Much turns on the force of Custer's written orders, which, judged by usual military documents of the kind, certainly gave Custer a much larger liberty than Colonel Hughes supposed. There is an affidavit of a witness who heard Terry's and Custer's last conversation together at the mouth of the Rosebud, just before Custer began his fatal ride. Terry said: 'Use your own judgment and do what you think best if you strike the trail; and whatever you do, Custer, hold on to your wounded.'"

General Miles says, in his book:

"But we have positive evidence in the form of an affidavit of the last witness who heard the two officers in conversation together on the night before their commands separated, and it is conclusive on the point at issue. This evidence is that General Terry returned to General Custer's tent,* after giving him the final order, to say to him that on coming up to the Indians he would have to use his own discretion and do what he thought best. This conversation occurred at the mouth of the Rosebud, and the exact words of General Terry, as quoted by the witness, are:

"'Custer, I do not know what to say for the last.'

"Custer replied: 'Say what you want to say.'

"Terry then said: 'Use your own judgment, and do what you think best if you strike the trail; and, whatever you do, Custer, hold on to your wounded.'

"This was a most reasonable conversation for the two officers under the circumstances. One had won great distinction as a general in the

* These two authorities seem to differ as to just when the conversation took place. Andrews, apparently quoting Miles, says: "Just before Custer began his fatal ride." Miles, quoting the mysterious and unknown affiant, says the conversation took place the night before, and at Custer's tent. The difference is radical and, in view of Colonel Godfrey's suggestion below, is material. Besides, the regiment marched away at noon on the 22d, and that is the date of the order; hence, Custer had no orders the night before. The regiment passed Generals Terry, Gibbon, and Custer in review as it marched away. When, then, was the precise hour at which this alleged conversation took place? — C. T. B.

Civil War; was an able lawyer and department commander, yet entirely without experience in Indian campaigns. The other had won great distinction as one of the most gallant and skilful division commanders of cavalry during the war, commanding one of the most successful divisions of mounted troops; he had years of experience on the plains and in handling troops in that remote country, and he had fought several sharp engagements with hostile Indians."

If General Terry spoke such words to General Custer the last thing before Custer's departure, those remarks have a very important, almost a decisive, bearing on the matter at issue. The only question then existing would be, how far the verbal order ought to be considered as superseding the written one. *It is my opinion that the charge that Custer disobeyed orders would fall to the ground if the truth of the alleged remarks could be established.* By giving him this verbal order, Terry would make Custer an absolutely free agent. The vital importance of establishing this affidavit is therefore obvious.

I call attention to the fact that Terry nowhere refers to this conversation, which it would be incumbent upon a gentleman to declare immediately Custer was charged with disobeying Terry's written order, and that Terry, in that portion of his report which is quoted by me on page 225, virtually not only fails to exculpate but actually charges that Custer did disobey his order, by saying he did the very thing that he was not expected to do.

To establish this affidavit, I wrote to President Andrews, asking his authority for stating that such an affidavit existed and requesting a copy of it. Here is his reply. I insert it without comment.

The University of Nebraska, Chancellor's Office,
Lincoln, November 22d, '03.

My Dear Sir:

Replying to your esteemed favor of the eleventh inst. I regret to say that I have no means of recalling with certainty the source of my information touching the Custer affidavit. My impression is, however, that my informant was Gen. Miles, with whom I communicated on the subject while I was writing my account. I also conversed personally with Hughes and with a very intimate friend, now deceased, of Gen. Terry's.

I shall be extremely pleased to read your views upon this subject.

Very truly yours,

E. Benj. Andrews.

I also wrote to General Miles and received the following reply from him:

1736 N Street, N. W.,
Washington, D. C., November 20, 1903.

My Dear Sir:

In reply to your two letters, you will find in my book, "Personal Recollections, or from New England to the Golden Gate," published by Werner & Co., Akron, Ohio, perhaps all the information you will require. I can not give the time now to going over the campaign in detail. I presume you will find the book in most libraries.* You will notice in it a chapter on the Custer campaign. General Custer did not disobey orders. When General Terry divided his command, taking one portion of it with him up the Yellowstone, and sending General Custer with the other portion far out in the Indian country, it necessarily put from seventy-five to one hundred miles between the two commands, and therefore placed upon General Custer the responsibility of acting on the offensive or defensive, for he could have been attacked by the whole body of the combined tribes, and, on the other hand, if he allowed them to escape without attacking them, he would have been severely censured. It would be silly to suppose that Indian chiefs like Sitting Bull and Crazy Horse would permit two columns to march around over the country with infantry, cavalry, wagon trains, etc.,† and wait for them to come up on both sides simultaneously, and one must believe the American people very gullible if they thought such a proposition had military merit.

Yours very truly,

NELSON A. MILES.

I immediately wrote General Miles a second letter asking him for the name of the affiant and any statement he might be willing to make about the affidavit. I pointed out to him what he very well knew — the prominence given to the story in his own book indicates that — the importance of the affidavit in establishing General Custer's position and defending him against the charge of disobedience. I received no answer to this letter.

Meanwhile the question of the affidavit was taken up by General Hughes in his several communications to me which appear below.

In order not to break the thread of the discourse I will anticipate

* I have it in my own library, of course, and have consulted it frequently. — C. T. B.

† This is overdrawn. Custer had neither infantry, artillery nor wagons with him; Gibbon had cavalry, infantry and artillery, but no wagons, be it remembered. — C. T. B.

events and here insert a third letter which I wrote to General Miles, after carefully considering General Hughes' remarks. The letter was sent to General Miles by registered mail. I hold the registry receipt showing that he received it. To this inquiry I have as yet received no reply.

455 East 17th Street, Flatbush,
Brooklyn, N. Y., March 30th, 1904.

Lieutenant-General NELSON A. MILES, U. S. A.,
1736 N Street, N. W., Washington, D. C.

My Dear General Miles:

A few months since I addressed to you a letter asking you for the name of the person, alluded to in your book, who made the affidavit as to the last remark of General Terry to General Custer. This letter has probably never reached you since I have never received any answer to it.

The statement is questioned by a number of officers, and in the interest of historical accuracy and for the sake of bringing forward every particle of evidence tending to clear General Custer of the charges which are made against him in that campaign, I most respectfully ask you to give me the name of the affiant together with such other statements concerning the affidavit as may be conclusive. How did you become possessed of the affidavit, for instance? Did you see it? Did you know the affiant? Was he a person whose testimony was to be implicitly relied upon? Is he alive now? In short, any information concerning it will be most acceptable as well as most useful.

Very sincerely yours,
CYRUS TOWNSEND BRADY.

I do not desire to comment on General Miles' refusal further than to say that if he has in his possession the affidavit he should either submit it to the inspection of impartial observers, give it to historians, state who made it, where it was made, furnish a certified copy of it to the public, or otherwise establish it. If he is not willing to do this he should at least say why he is not willing. I submit that no man, whatever his rank or station, ought to make statements which affect the fame and reputation of another man *without giving the fullest publicity to his sources of information*, or *stating why the public must be content with a simple reference thereto.*

While I am on the subject of the affidavit, I call the student's attention to a possible suggestion in Colonel Godfrey's second communication below.

It is twenty-eight years since the Battle of the Little Big Horn. If the alleged affiant is now alive, what reason can exist to prevent him coming out and acknowledging his affidavit? If he is dead, why should secrecy about it longer exist? Why does not General Miles break his silence? *The whole matter turns on the production of this affidavit, with satisfactory evidence as to the character of the affiant.*

The other position taken in General Miles' letter above, which of course is a summary of his views as set forth in his book, is discussed later on by General Woodruff.

II.

I now refer the student to the following letter in answer to one from me asking information and calling General Hughes' attention to President Andrews' book, which has just been reissued in a new and amplified edition:

New Haven, Conn.,
18th Nov., 1903.

Dear Sir:

Your letter of the 13th was duly received. I had not heard of Dr. E. Benjamin Andrews' book prior to receipt of your letter, but have looked it up since.

After a careful examination of what he says about the Sioux campaign of 1876, I cannot find any good and sufficient reason for changing aught that was stated in the article published in the *Journal of the Military Service Institution*, in January, 1896. I do find, however, that something could be added to the statement of the case in reply to new matter which he has injected into it in his book. These items are three in number, to wit:

1st. General Miles does not agree with the views therein expressed.

2d. New evidence in the form of an affidavit made by some individual, name not given.

3rd. The writer of the book dissents from my view of the case.

We will take these items up severally:

First: "General Miles is strongly of the opinion that Custer was not guilty of disobeying any orders."

It is not a new experience to learn that the views of General Miles and myself are at variance. Indeed, it seems that they are seldom in accord. But, in this instance, my views are supported by the late General P. H. Sheridan, who states as follows:

"General Terry, now pretty well informed of the locality of the Indians, directed Lieutenant-Colonel George A. Custer to move with the Seventh Cavalry up the Rosebud, until he struck the trail discovered by Major Reno, with instructions *that he should not follow it directly to the Little Big Horn, but that he should send scouts over it and keep his main force farther south.*" *

General Gibbon, in a letter to General Terry, written after having reached his post, Fort Shaw, Montana, and bearing date November 5th, 1876, writes as follows, speaking of the "Conference":

"We both impressed upon him (Custer) that he should keep constantly feeling to his left, and even should the trail turn toward the Little Big Horn that he should continue his march southward along the headwaters of the Tongue, and strike west toward the Little Big Horn. So strong was the impression upon my mind and great my fear that Custer's zeal would carry him forward too rapidly, that the last thing I said to him when bidding him good-by, after his regiment had filed past you when starting on the march, was, 'Now, Custer, don't be greedy, but wait for us.' Poor fellow! Knowing what we do now, and what an effect a fresh Indian trail seemed to have on him, perhaps we were expecting too much to anticipate a forbearance on his part which would have rendered coöperation between the two columns practicable."

The foregoing clearly shows that no doubt existed in the minds of the Division Commanders and the third party present at the conference as to what the instructions required and that those instructions were not complied with.

Second: Dr. Andrews states that there was a listener at the last conversation between Terry and Custer at the mouth of the Rosebud, just before Custer began his fatal ride, and that his affidavit sets up that:

"Terry said: 'Use your own judgment and do what you think best if you strike the trail and, whatever you do, Custer, hold on to your wounded.'"

It is quite evident that this is the same affidavit which General Miles refers to in his book. My attention was called to that reference in the winter of 1896, and in behalf of the family and friends of General Terry I asked to see the affidavit, saying that I might wish to make a copy of it. My request was refused by General Miles, with

* Report of Secretary of War, Vol. I., 1876, page 443. Italics in quotation above are mine. — C. T. B.

the further information that it had been in his possession for nineteen years, which carried the date back to a time when Colonel Miles was commanding a post in General Terry's Department. The value of the document could have been very readily determined at that time by General Terry himself, and I am thankful to say its worthlessness is still capable of proof.

The quotations given by Dr. Andrews would alone be sufficient to condemn the paper with any one familiar with General Terry and the situation.

It will be noticed that he is represented as saying, "If you strike the trail." Terry was sending, with Custer, Reno and six troops of cavalry, who had followed the trail for many weary miles only three days before, and there was no "if" in the case.

The other quotation is equally incredible. General Terry had an enviable reputation throughout the army for his exceptional courtesy on all occasions and under all circumstances, to all those serving in his command. To have made the remarks quoted, "Whatever you do, hold on to your wounded," would have been tantamount to saying to one of his Lieutenant-Colonels, to whom he was confiding the finest command in his department, that he considered it necessary to caution him on the elementary principle of the position assigned him. To an officer of General Custer's experience and gallantry such a caution would have been far from agreeable, and such action would have been entirely foreign to the life-long conduct of General Terry.

But, fortunately, we are not dependent upon deductive evidence in this instance. There are still living a good many people who were witnesses of that "march past" and parting of Generals Terry and Custer. By personal observation I positively know that any one, General Gibbon excepted, who makes affidavit to the effect that, at the said parting at the Rosebud, there was a conversation between Terry and Custer to which he was the only listener, is guilty of perjury.

When the notice came that the command was ready to take up its march I was sitting with Terry and Gibbon. General Terry invited General Gibbon to go with him and see it. They walked off a few yards from the bank of the stream and stood together when Custer joined them. The three remained together until the command had filed past and the final good-by was said. Custer mounted his horse and rode off, and Terry and Gibbon came back to where I had remained. The last remark made to Custer was by Gibbon. If any change was made at that time in Custer's orders it was perforce known to Gibbon, who was alongside of Terry, and the only one who was with him and Custer. Now let us see what Gibbon said in a letter written to General Terry twenty-seven years ago, when he could not foresee for what purpose his words would be quoted:

"Except so far as to draw profit from past experience it is, perhaps, useless to speculate as to what would have been the result had *your plan* been carried out. But I cannot help reflecting that, in that case, my column, supposing the Indian camp to have remained where it was when Custer struck it, would have been the first to reach it; that with our infantry and Gatling guns we should have been able to take care of ourselves, even though numbering only two-thirds of Custer's force."

The only person actually in the presence of Terry and Custer at that final parting, happened to be the third member of the conference, who knew the "plan," and on the fifth of the November following he writes as above, of what would have been the result had "your plan" (Terry's conference plan) been carried out.

It so happened that I went over this whole subject with General Gibbon personally only a short time before his death. He certainly knew nothing of any change in the "conference plan" at that time. Any historian who makes use of the affidavit General Miles had some years ago, would do well to look carefully into the facts.

Third: Andrews states: "He (Hughes) adduced many interesting considerations, but seemed to the writer not at all to justify his views."

I am, by no means, sure that this dissent would have disturbed me if I had depended on my own judgment alone in submitting the article* for publication, but it so happened that I did not do so. The tragedy discussed being of an exceedingly grave character, and both the responsible heads having passed away, rendered it unusually important that every possible precaution should be taken against mistakes. For this reason, after the "many interesting considerations" were prepared for publication, I submitted the article to different *competent military men* with the simple question, "Is it conclusive?"

The final review was made by General Henry L. Abbott, U. S. Army, who enjoys a world-wide reputation for military ability and scholarly attainments. The gentlemen who revised the article were unanimously of the opinion that the statement was *absolutely conclusive,*

* The reference is to the article in the *Journal of the Military Service Institution* mentioned above.—C. T. B.

and with the support of such men I have felt little anxiety about the criticisms that have been made.

Very truly yours,

R. P. Hughes.

Thereafter I wrote again to General Hughes about some matters repeated to me in conversation by General Carrington, who told me that Custer actually got down on his knees to Terry and begged him, for the sake of Custer's honor and fame as a soldier, to get the orders detaining him at Fort Lincoln revoked, so that he might be spared the disgrace of seeing his regiment march to the front leaving him behind. Carrington's recollection was that the scene took place in Terry's bedroom.

Here is General Hughes' letter on that point:

New Haven, Conn.,
27th Nov., 1903.

My Dear Sir:

Yours of the 27th at hand. Carrington is all right except as to location—the incident occurred in General Terry's office in St. Paul, corner Fourth and Wabashaw Streets. It drew from Terry a request to the President to permit Custer to go with him, the answer being through Sherman, "If Terry wishes Custer let him take him along." Just after notifying Custer of the reply and telling him he would take him along, Custer met Ludlow on the street and made the "swing clear" remark which is spoken of in Andrews' history and is referred to in my article in the journal. I shall have to apply for a copy of the Secretary of War's report for 1876, which has the reports of Sheridan, Crook, Terry, Gibbon, Reno, etc., pages 439-487.*

Yours very truly,

R. P. Hughes.

III.

Meanwhile I had communicated with Colonel Godfrey, who had already furnished me with much data in addition to that contained in his valuable and interesting paper, calling particular attention to some of the statements made by General Hughes in his article in the *Journal*

* I have had them before me constantly for the past six months, and have examined them most carefully again and again, verifying quotations, etc.—C. T. B.

of the Military Service Institution. From Colonel Godfrey I received the following paper:

MEMORANDA FOR REV. C. T. BRADY

A semi-official account entitled "Record of Engagements with Hostile Indians in the Division of the Missouri, from 1868 to 1882," was published by the Division of the Missouri. This paper is now being reprinted in the *United States Cavalry Journal*, Fort Leavenworth, Kansas. The part relating to the Little Big Horn, is in the October, 1903, number. This account reads:

"About two o'clock in the morning of July 25th, the column halted for about three hours, made coffee, and then resumed the march, crossed the divide, and by *eight o'clock* were in the valley of one of the branches of the Little Big Horn."

This is misleading and not altogether true. We halted about two A.M., till eight A.M., then marched till ten A.M., halted, and it was not until nearly noon that we crossed the divide. We were in a narrow valley. The march is correctly described in my article. The inference is that Custer was so very eager that he crossed the divide into the valley of the Little Big Horn and put himself where he could be discovered. General Hughes' article is a *special plea* to clear General Terry from the odium that he and his family seemed to think was heaped upon him for failure to push forward with the information they had on June 25th and 26th, and that General Custer's family and friends were supposed to hold him (Terry) responsible for the disaster in a measure. I do not remember a charge of disobedience as having been made at any time during this campaign; nor, on the other hand, do I recall that much was said that Terry and Gibbon did not do as they thought best on June 25th and 26th.

The marching distance from the Yellowstone, where Gibbon's command was crossed, to the Little Big Horn, was about forty-six miles. East of the Big Horn, the country over which Gibbon's forces marched, was rough—bad lands. The Second Cavalry on its march June 25th, saw the "big smoke" (from the fire in the bottom at the time of Reno's attack) and at once sent word to General Gibbon (and Terry) that they thought a fight was going on, or something to that effect. I don't know when they saw this "big smoke," but my recollection is that it corresponded or tallied very well with the time of Reno's attack. The Second Cavalry got to the Little Big Horn, four or five miles above the mouth, about nine-thirty, June 26th. They were then distant from the battlefield about eight or ten miles — an infantry officer says six miles. They arrived in the vicinity of our position about eleven A.M., June 27th, nearly two days after the "signs," the "big smoke" of the fight, had been communicated.

That the country between the Yellowstone and the Little Big Horn was rough; that the 25th of June was hot; that the water was scarce, we all know; but we thought it strange that, after they learned from the Crow scouts — say at ten-thirty, June 26th, on the Little Big Horn — that a disaster had occurred, it took them so long to get a move. Yet none of us blamed them for being cautious at that time.

General Terry was not an Indian fighter and would never have made a success of getting Indians on the plains. The idea is preposterous * that a force can march through the open country (a great big country like we had) pass by the Indians fifty or sixty miles south, then turn round and find them in the same place, and crush them between that force and another from the opposite direction. They don't linger that way. Our march from eleven P.M., June 24th, was in a close country and not exposed — was in a close valley, a branch of the Rosebud. The Indians who discovered us and sent word to the village would have discovered our trail and consequently informed them of our movements.

General Custer did not intend to attack until June 26th, the date Terry was to be at the Little Big Horn.† Herandeen was the scout that was to take the information through to Terry, but developments made it necessary, in General Custer's opinion, if we were to strike the Indians at all, that we should do it at once. Even then he expected only a running fight. Their stand and concentration were unexpected, because the chance of "surprise" was gone, and he probably did not send Herandeen, as was intended, to communicate with Terry for the reason that he did not think he could get through.

Now, suppose the Indians had been located on the headwaters of the Rosebud or Tongue, or Powder, and not on the Little Big Horn, and we had bumped up behind them on the north, should Custer have backed away, sent a scout through to Terry, made a detour so as to get

* If the orders were preposterous, or involved movements that were profitless and absurd, why did not Custer point out these patent absurdities to Terry and Gibbon *before he started?* There had been no change in conditions; the trail, the Indian position, and everything else were just as the orders predicted.— C. T. B.

† The time of the arrival of Terry at the Little Big Horn is assumed to be June 26th. What authority there is for that assumption I do not now recall. It is not embodied in the "instructions." We of the command knew nothing of it till after the battle; after Terry's arrival, that is.—E. S. GODFREY.

to the south side? Terry's instructions had fairly located the Indians, but it was a mere *guess*.* On the 17th they had fought Crook to a retreat, then they concentrated upon the Little Big Horn.

In my opinion, if our attack had been delayed even a few hours we would not have found the Indians all in the village. When we got to the divide their pony herds were still out grazing; when the attack was made all herds had been driven into the village; they did not have time to strike their tepees and steal away. I don't believe they had a long warning of our advance. The Indian runners had the same, or a greater, distance to get back than we had to advance. It was their evident purpose to drop out of sight of our scouts who were in position for observation before daylight, and did not see them returning down the valley on the trail. Therefore, they must have made a wide detour.

Again, when they discovered us we were probably in bivouac and, at all events, an ordinary day's march distant from the village. The time of warning, I think, could be safely conjectured as the time of arrival of the few warriors that came out to meet the advance and attack Reno. All those warriors that had their ponies handy, I believe, were assembled at once to come out and meet the troops. The rapid advance didn't give the Indians a chance to collect their belongings and mature any plans to escape; otherwise I believe the expected "scattering" would have taken place. And in just so much was the attack a "surprise."

That General Custer deliberately disobeyed Terry's orders I do not believe. Custer was intensely in earnest and fully determined to *find* the Indians and, when found, to attack them, even if it took him back to the agencies. Suppose Custer had asked Terry "If I find these Indians shall I attack, or wait for you?" Undoubtedly Terry would have replied "Attack!" He was too good a soldier not to appreciate opportunity, but he was not enough of a cavalryman or Indian fighter to appreciate the flash-like opportunities for hitting the Indians on the broad prairies.

Custer was what in these modern days is styled a "strenuous" man. Terry was not. He was the personification of gentleness and deliberateness. And besides, Terry's instructions *gave the necessary latitude*. He told Custer what he thought should be done but, after all, left it to Custer's judgment and discretion when so nearly in contact with the

* Having located them, guess or not, the conditions were exactly those contemplated in the orders.—C. T. B.

enemy.* If Custer had passed on south and the Indians had escaped, or had gone forth and attacked him, as they had Crook, and defeated him, would these instructions have shielded him? Not much. He would have been damned as cordially for the failure of the expedition as he is now, by those same men, for courting disaster. I have no doubt in my mind, that if Custer had passed south even one more day, the Indians would have attacked us as they had General Crook, and upon almost the same ground, just one week before.†

Terry says, in his instructions, "He will indicate to you his own views of what your action should be, and he desires that you should conform to them unless *you should see sufficient reason for departing from them.*" ‡ Custer was an experienced war soldier, a thorough cavalryman, and an experienced Indian campaigner. So why not give him the benefit of "sufficient reason"? Were Terry's instructions "definite and explicit"? Terry himself says in his order that "definite instructions" were "impossible."

There was not an officer or soldier of the Seventh Cavalry but that expected a fight when we were preparing to leave the mouth of the Rosebud. Where the fight would take place we knew not, but I venture to say that never was there a thought that the Indians would take a position and wait there for us to go through a lot of manœuvers. Reno's scout had not brought any *definite* information. I find my notes (June 20th) say that it was generally thought the trail, when they left it, was about *three weeks old* and the indications showed perhaps three hundred and fifty lodges. I don't think General Terry had any later information than Reno's scout on which to guess the location of the Indians on the Little Big Horn. General Custer's statement that he would follow the trail until he found the Indians, even if it took us to the agencies on the Missouri or in Nebraska, does not indicate that he expected them to wait in position on the Little Big Horn or elsewhere. This statement was made after it had been decided that we should go over the trail, June 21st, but probably before the general instructions had been made out.

As it turned out I think Custer did make a mistake in going in with a divided force, not that the division of itself would have been fatal, but because Reno failed to hold a leg even if he couldn't skin.

If Custer had followed Reno the latter, in my opinion, would never

* But if Custer had followed his orders, he would not have been nearly in contact with the enemy — there's the rub!—C. T. B.

† This I consider a good point in Custer's favor.—C. T. B.

‡ Italics mine.—E. S. Godfrey.

have dared to halt, or even hesitate, in his attack. If Reno had even held to the bottom, the overwhelming forces would have been divided. There was nothing in Reno's past career that would indicate confidence should not be placed in his courage. Custer could not have anticipated a faint-hearted attack or that Reno would get stampeded.

I believe that Reno was dismayed when he saw the showing in front of him, and when he failed to see the "support" promised, I think he lost his nerve, and then when his Ree scouts stampeded and he found his force being surrounded in the bottom, I believe he abandoned himself to his fears, then stampeded to the hills and lost his reason, throwing away his ivory handled pistols. If Reno had held to the bottom, Custer's left flank (Keogh and Calhoun) would not have been so quickly overwhelmed (for the Indians leaving Reno made that envelopment), and it is reasonable to suppose Custer would have had a better show to withdraw and rejoin other forces.

If Custer had followed up Reno he would have taken matters in his own hands, held and concentrated his men in such manner as to control the situation until Benteen and the packs came up. The Indians, as a rule, will not stand punishment unless cornered. I went over the ground in the bottom where Reno was when he concluded to go to the hills, and I believe he could have held the position. I talked the matter over with General Gibbon and he practically agreed with me. I know many others think otherwise, including some who were in that part of the fight.

I have doubts about the saving of Custer if Reno had advanced after the packs joined us, for I think the fight was practically over then. To have advanced before then might possibly have done something in favor of Custer, but probably not. I am of the opinion that part of the fight was settled quickly. Custer's battalion had practically no shelter and no time to make any. While a good many horses were killed, I fear that most of those getting away carried their reserve ammunition, and it didn't take long to get away with fifty rounds in a fight. With a different commander than Reno we might have created a diversion by advancing as soon as the ammunition packs came up, which was some little time before McDougall arrived with all the packs. Reno was apparently too busy waiting for further orders from Custer to take any initiative. Weir asked permission to take his troop to reconnoiter in the direction of the firing on Custer, and Reno would not give it. Weir started on his own hook, and Edgerly (Weir's Lieutenant) supposing permission had been given for the troop, followed Weir with the troop. I think Reno subsequently tried to make it appear that this advance of Weir was by authority. I don't think Reno was drunk, for I don't believe there was enough whiskey in the command to make a "drunk." ‡

At the Reno Court of Inquiry I was asked if I thought Reno had done all he could as a commanding officer, and I replied "No." That was about the effect of the question and answer. The testimony and proceedings were reported in full in the Chicago *Times*. The New York *Herald* had an able correspondent, Mr. Kelly, that joined our forces on the Yellowstone in July or August, and wrote, giving all the information he could gather from all sources that pervaded the command, that he could *get at*. There were a "whole lot" of correspondents in the field after the fight, but Mr. Kelly was considered one of the ablest. Being in the field till September 26th, we saw but few newspapers from the east.

E. S. Godfrey.

On the receipt of this memoranda I sent Colonel Godfrey all the papers printed above, and asked him further to discuss these papers. They were returned to me with the following letter, accompanied by these additional notes:

Headquarters, Ninth United States Cavalry,
Fort Walla Walla, Washington,
February 12th, 1904.

My Dear Doctor:

I return to-day the letters sent to me by registered mail. I am very sorry to have kept them so long from you, but I have been suffering from a sprained knee which has laid me up, and have been otherwise under the weather.

I feel that I have not in my memoranda done justice to the subject. It is largely one of sentiment, and the best rule is to put yourself in his place and act under the lights then exposed to view. That Custer may have been actuated by other motives I do not doubt. The main question to me was whether he was justified from a military point, in a campaign against Indians, in his conduct of the march and battle.

If we could have foreseen as we now look back and see!

Sincerely yours,

E. S. Godfrey.

‡ Colonel Godfrey made this statement in answer to a question from me. On this point see Appendix B.—C. T. B.

ADDITIONAL NOTES BY COLONEL GODFREY.

The statement of General Sheridan, quoted by Hughes, was made in his annual report for 1876, and of course from data furnished by General Terry. It is but natural that he should reflect more or less the views of Terry. He could have had only the newspaper and other unofficial accounts. Of course I recognize that "unofficial accounts" very often give more *inside* information than the official report.

A word as to that affidavit. I don't know anything about it and am ready to take Hughes' say-so as to what *officers* were present, but I suggest a possible solution: When Custer dismounted he had his orderly and generally his flags with him; naturally the orderly would be somewhat retired, and when Custer went to mount his horse Terry may have gone aside to accompany him and spoken the caution to him in a subdued voice so that Gibbon would not have heard him, but the orderly might have heard.*

In going over a lot of letters relating to the campaign, etc., I find one from General J. S. Brisbin (now dead), then Major, commanding Second Cavalry Battalion. It is dated January 1st, 1892, just two weeks before his death. In it he is very bitter against Custer. He says that Custer disobeyed:

"If not in letter, then in spirit, and I think and have ever thought, in letter as well as spirit. Terry intended, if he intended anything, that we should be in the battle with you. I was on the boat, steamer Far West, Captain Grant Marsh, the night of the 21st, when the conference took place between Gibbon, Custer and Terry, to which you refer, and I heard what passed. Terry had a map and Custer's line of march up the Rosebud was blocked out on it by pins stuck in the table through the paper. Terry showed Custer his line of march and, being somewhat near-sighted as you know, Terry asked me to mark the line, and I did so with a blue pencil. Custer turned off that line of march from the Rosebud, just twenty miles *short* of the end of the pins and blue line."

Just how much dependence can be placed on Brisbin's statements I don't know. He may have been present at this conference, but Hughes

* Again I ask General Miles if this is the explanation of the affidavit? If so, how does it accord with the statement that the conversation occurred the night before the command separated? Or, has General Miles written carelessly "the night before," and does he mean just before the final march past? — C. T. B.

makes no mention of him; in fact, entirely ignores him and may have forgotten him. I will make another quotation from Brisbin :

"I read the order you print as being the one given by Terry to Custer for this march. If that is the order Custer got it is not the order copied in Terry's books at Department Headquarters. You will remember that after Custer fell Terry appointed me chief of cavalry. I looked over all the papers affecting the march and battle of Little Big Horn and took a copy of the order sending you up the Rosebud. The order now lies before me and it says 'you should proceed up the Rosebud until you ascertain definitely the direction in which the trail above spoken of leads (Terry had already referred to the trail Reno followed). Should it be found, as it appears almost certain that it should be found, to turn toward the Little Big Horn, he thinks (that is, the Department Commander thinks) that you should still proceed southward, perhaps as far as the headwaters of the Tongue River, and *then* ('then' underscored in order) turn *toward Little* Big Horn, feeling constantly, however, to your left, so as to preclude the possibility of the escape of the Indians to the south or southeast by passing around your left flank. It is desired that you conform as nearly as possible to those instructions and that you do not depart from them unless you shall see absolute necessity for doing so.'"

That part of the quotation from "It is desired" to "necessity for doing so," is omitted in the order as printed in the report of General Terry. Not having seen the original order I cannot vouch for either being the true copy, but the omission looks peculiar to say the least, if omission there was.*

I do not know that I can add very much to what I have already sent to you on the question of disobedience. Here is a commander who has had experience in war, civilized and Indian, sent in command of his regiment against an unnumbered foe, located we know not where (although well conjectured in the instructions, as it turned out); given instructions to *preclude their* escape; to coöperate with another column separated from fifty to one hundred miles, having infantry and artillery, marching over a rough, untried country. Now if that commander thought that to go on farther south before he had located the foe (when he was on the trail) was to leave an opening and an almost cer-

* Personally I do not believe that the sentence in question was in the order given to Custer. For if it was, why should Terry suppress it, since it only confirms his own claims? Besides I should be loath to believe that Terry would suppress anything. The sentence may have been in a rough draft of the orders, and not in the final copy.— C. T. B.

tainty of their escape, if they wanted to do so, is it reasonable to expect him to leave the trail and go on "in the air"? The commander who gives him his instructions cannot be communicated with. Is this isolated commander not allowed to act on his own responsibility, if he thinks he cannot *preclude* the escape by leaving the very trail that will locate the enemy?

Hughes in his article, and the official reports, make it appear that we were at or near the "Crow's Nest" at daylight and crossed the divide at eight A.M. The scouts were at the "Crow's Nest," but at eight A.M. we took up the march to near the divide and "Crow's Nest," arriving at ten o'clock, A.M.; that is, we were in the Rosebud Valley, one mile from the divide. We did not cross the divide till nearly noon. Hughes seems to pooh-pooh the idea that we were not to attack till the morning of the 26th. We had Custer's own statement as to that. He said so himself when he called the officers together on the night of June 24th and again reiterated the statement before crossing the divide.

During the second or third day (23rd or 24th) up the Rosebud, several times we thought we (I mean some of us) saw smoke in the direction of the Tullock, and finally we spoke of it to the General (Custer) at one of the halts. He said it could not be, that he had scouts over on that side and they most certainly would have seen any such "signs" and report to him, and he reiterated that there were scouts out looking toward Tullock's Valley. After this assurance we made it a point to watch this "smoke business" and we discovered the illusion was due to fleecy clouds on the horizon and the mirage, or heated air, rising from the hills on that side. The air was full of dust from our marching columns, which helped the illusion.

With reference to my slip that "about eighteen hundred had gone *from one agency* alone." I took that from my diary, as I had been informed by some one who got the information from Department Headquarters. I had never seen the despatch and put down the item as it came to me. It was a matter of common report in the camp.

Another point occurs to me: "For Custer to be in coöperating distance on the only line of retreat if the Indians should run away." (Hughes' magazine article, page 36.) Hughes intimates that there was only one line of retreat, presumably up the valley of the Little Big Horn. The Indians certainly could have retreated over their traveled route, or could have cut across the headwaters of the Tullock for the Yellowstone had Custer gone south. Hughes seems to forget that an almost impassably rough country — the Wolf Mountains — would lie between Custer and those lines of retreat. Yet he would insist that it was good generalship to leave these routes open to close up one other. The Indians were in light marching order and could travel faster than Gibbon over the Tullock Divide, and there would have been a long-distance, "tail-end" pursuit for Custer when he descended the Little Big Horn (by following the "plan") and found the enemy had escaped over the very trail he had left behind him, or had struck for the Yellowstone, passing Gibbon's left.

It has been the criticism almost ever since Indian fighting began that commanders were too prone to follow some strategic theory and fail to bring the Indians to battle — give them a chance to escape. It was Custer's practice to take the trail and follow it, locate the enemy and then strike home by a surprise attack. Custer knew the ridicule and contempt heaped on commanders who had failed to strike when near the enemy; or who had given the enemy opportunity to escape when nearly in contact with them. Whatever may be the academic discussions as to his disobedience, I hold that he was justified by sound military judgment in making his line of march on the trail.*

IV.

General Hughes and Colonel Godfrey may be considered fairly enough as representatives of the opposing views on the question. I thought it would be well to have the papers discussed by an officer who might be considered as taking an impartial view of the matter. I therefore sent them to Brigadier-General Charles A. Woodruff, U. S. A. (retired), and his review of the whole question is as follows:

103 Market Street,
San Francisco, May 3, 1904.

My Dear Dr. Brady:

I have read with a great deal of pleasure, your three articles on "War with the Sioux," and I have taken the liberty of making various marginal notes and corrections on the manuscript. I have also read the letters from General Miles, Professor Andrews, General Hughes, and Colonel Godfrey.

General Miles, in his letter of November 20, 1903, dismisses the matter very curtly. He says "Custer did not disobey orders," and he

* This also is very interesting and seems to point to the order as a "preposterous" one under the circumstances. It may be so; but if so, I wish Custer had pointed it out to Terry before he started. — C. T. B.

states as military dictum that, in sending General Custer seventy-five or one hundred miles away, Terry could not indicate what Custer should do, and that, practically, Custer was not under any obligations to execute Terry's orders, even when he found conditions as Terry had expected and indicated.*

The order states explicitly "Should it — the trail up the Rosebud — be found (as it appears almost certain it will be found) to turn toward the Little Big Horn, then you should still proceed southward." Now, when he found that it turned toward the Little Big Horn, instead of going south or stopping where he was and scouting south or southwest and west and try to locate the village, or examining Tullock Creek, or sending scouts to Gibbon, he made that fatal night march with the deliberate intention of trying to locate and strike the village before Gibbon could possibly get up.

Gibbon says (page 473, Vol. I., Report of the Secretary of War for 1876), "The Department Commander (Terry) strongly impressed upon him (Custer) the propriety of not pressing his march too rapidly." Whether Custer's written instructions were based upon a "guess" of the actual condition, as Colonel Godfrey suggests, or had no "military merit," as General Miles states, the facts remain: First: That they were based upon a "foresight" as good as the present "hindsight," which is often not the case. Second: That Custer accepted them without demur. Third: No further information was gained to suggest a modification, or, to use the words of the letter: "unless you see sufficient reason for departing from them." On the contrary, the supposed turn of the trail was found to be an actual fact.

Therefore, Custer did not obey his written instructions, in letter or spirit, and had no proper military justification for not doing so, unless General Terry afterwards told him, "Use your own judgment and do what you think best," which, in my opinion, would have made the instructions advisory rather than positive orders. If these facts (I ignore the unproduced affidavit) do not constitute disobedience of orders, I do not see how it is possible for the charge of disobedience of orders to hold against any man, under any circumstances, when away from his superior.

Here is a trifling sidelight on the matter. On the night of June 23d, General Gibbon, in reply to an optimistic remark of mine, told me in effect, "I am satisfied that if Custer can prevent it we will not get into the fight." The meaning I gathered was that Gibbon thought that

* Would General Miles excuse subordinates for such obedience (?) of his orders for a combined movement? — C. A. WOODRUFF.

Custer was so eager to retrieve the good opinion that he might have lost owing to his controversy over post traderships, that he would strike when and where he could.

While Terry, with Gibbon's command, was camped at Tullock's Creek, Saturday night and Sunday morning, June 24th and 25th, he was looking for a message from Custer very anxiously, so I was told at the time.

Colonel Godfrey speaks of the odium Terry's family seemed to think was "heaped upon him for the failure to push forward on the information they had on June 25th and 26th." Now let me say a few words with reference to that.

The smoke that is spoken of as having been seen by Terry's command — and I saw it myself — was on the afternoon of June 25th. It was occasioned, I understood, by attempts to drive some of Reno's stragglers out of the brush, and must have been somewhere from two to four o'clock in the afternoon. Now let me quote from a telegram of General Terry, dated June 27, and found on page 463, Vol. I., Report of the Secretary of War of 1876, to show that Gibbon's command did not linger by the wayside:

"Starting soon after five o'clock in the morning of the 25th, the infantry made a march of twenty-two miles over the most difficult country which I have ever seen. In order that the scouts might be sent into the valley of the Little Big Horn, the cavalry, with the battery, was then pushed on thirteen or fourteen miles farther, reaching camp at midnight. The scouts were sent out at half-past four in the morning of the 26th. They soon discovered three Indians, who were at first supposed to be Sioux; but, when overtaken, they proved to be Crows, who had been with General Custer. They brought the first intelligence of the battle. Their story was not credited. It was supposed that some fighting, perhaps severe fighting, had taken place; but it was not believed that disaster could have overtaken so large a force as twelve companies of cavalry. The infantry, which had broken camp very early, soon came up, and the whole column entered and moved up the valley of the Little Big Horn."

I want to say that the infantry broke camp about four o'clock on the morning of the 26th. It had rained that preceding night and the lash ropes of the packs were soaked with water and, as we moved, they stretched continuously and we were stopping constantly to replace the packs, and besides that, mind you, traveling in adobe mud was very trying. I continue the quotation as follows:

"During the afternoon efforts were made to send scouts through to what was supposed to be General Custer's position, to obtain information of the condition of affairs; but those who were sent out were driven back by parties of Indians, who, in increasing numbers, were seen hovering in General Gibbon's front. At twenty minutes before nine o'clock in the evening, the infantry had marched between twenty-nine and thirty miles. The men were very weary and daylight was fading. The column was therefore halted for the night, at a point about eleven miles in a straight line from the mouth of the stream. This morning the movement was resumed, and, after a march of nine miles, Major Reno's intrenched position was reached."

It was the general opinion from indications found next day just beyond where we halted for the night, that had we proceeded five hundred yards more, we would have been in the midst of a night attack from the Sioux Indians, who came to meet us as a means of guarding their fleeing village.

In reference to the number of Indians, the same telegram of General Terry's says: "Major Reno and Captain Benteen, both of whom are officers of great experience, accustomed to seeing large masses of mounted men, estimate the number of Indians engaged at not less than twenty-four hundred. Other officers think that the number was greater than this. The village in the valley was about three miles in length and about a mile in width. Besides the lodges proper, a great number of temporary brushwood shelters were found in it, indicating that many men, besides its proper inhabitants, had gathered together there."

I am under the impression now that we counted positions occupied by twelve hundred lodges.

I coincide with your view that had Reno proceeded in his attack, with the audacity that should characterize, and usually does characterize, a cavalry charge, there would have been a different story to tell; perhaps as many men would have been killed, but they would have been divided among at least eight, if not eleven, troops of cavalry rather than concentrated in five, which meant annihilation for those.

I have been told, or was told at the time, that it was thought that about sixty lodges were a few miles up the Little Big Horn above the main village, and that, in the early morning, when Custer's proximity was discovered, that this small village, knowing that they were but a mouthful for Custer's command, hurriedly packed up and dashed down the valley. It can readily be understood that sixty lodges, with the horses and paraphernalia, moving rapidly down the valley, might well create the impression that a very large force was in retreat.

Now, if the Indian village was in retreat, Custer's division of his forces was not altogether bad. One command to hurry them up and continue the stampede, his main force to attack them in the right flank if they turned that way, which was most probable, Benteen's to attack them if they turned to the left, which was possible but not as probable.

Unfortunately for Custer they were not fleeing. Colonel Godfrey rather dwells upon the fact that Custer had to attack these Indians or they would have gotten away from him. The fact is, as I have stated above, when he left the Rosebud he did not know where they were, had not located them, was not in visual contact even with them, and a glance at the map will show that, standing on the Rosebud, where the trail left it to go over to the Little Big Horn, Custer was in the best possible position for intercepting these Indians on three of their four lines of retreat. For having passed into the Little Big Horn Valley, there were only four practicable routes of flight for the Indians, north, toward Gibbon, or east, northeast, or southeast. From the point where he left the Rosebud, Custer was in a position to strike either one of the three last lines of flight, whereas, if, after making the forced night march with his fatigued animals, he had struck the Little Big Horn, and a reconnaissance had shown that the village had left the Little Big Horn, going northeast, on the 24th of June, he would have been two days' march behind them.

Had he sent a scout, on the night of the 24th, to Gibbon, whose exact whereabouts was almost known to him, that scout would have reached Terry or Gibbon, on Tullock's Fork, a few miles from the Yellowstone, on the morning of Sunday, and by Sunday night Gibbon's command would have been within less than ten miles of what is designated as Custer Peak, the hill on which Custer perished. Then, with Custer moving on the morning of the 26th, Gibbon's infantry and Gatling guns could have forced those Sioux out of the village on to the open ground, extending from the Little Big Horn to the Big Horn, and Custer's twelve troops of cavalry and Gibbon's four, sixteen troops in all, between them would have made the biggest killing of Indians who needed killing ever made on the American continent since Cortez invaded Mexico. While this is a speculation, and an idle one, it is to my mind a rather interesting one.

I think myself that General Hughes makes out his case in reference to that affidavit that General Miles has so carefully treasured for so many years. It would be a very interesting historical document, but it would have been more satisfactory if it had been produced while Terry or General Gibbon or both were alive. I doubt very much whether

Major Brisbin's supposed copy of the order book at Terry's headquarters was compared with the original after Brisbin had made it.

I regret to say that my paper upon this campaign was lost, and I have not even the notes from which it was written. I found one brief page, which I quote merely as indication of my reasons for believing that there were more than two thousand Indian warriors in the battle of June 25th: "Before May 10th of '77 more than one thousand warriors came in and surrendered, not including the warriors killed in that battle or the half dozen other engagements, nor the individual warriors by the hundreds that sneaked back to the agencies and those who went to British America under Sitting Bull, numbering, it was understood, over two thousand warriors."

I do not think you are too severe upon Major Reno. I conversed with most of the officers of that command at one time or another, while in the field, and nearly all were very pronounced in their severe criticism of Reno. The testimony at the Reno court of inquiry was less severe than the sentiments expressed within a few days, weeks, and months after the occurrence. That was perhaps natural. It is barely possible that some of it was due to the fact that Captain Weir, one of General Custer's most pronounced friends and one of Major Reno's most bitter critics, died before the court of inquiry met.

I do not think that Sturgis, Porter, etc., were captured and tortured. I found most of the lining of Porter's coat in the camp, which showed that the bullet that struck him must have broken the back and passed in or out at the navel. My theory has been, with reference to those whose bodies could not be found, that most of them made a dash into the Bad Lands in the direction of the mouth of the Rosebud, where they had last seen General Gibbon's command. It would have been easy for them to have perished from thirst in the condition they were in, and if they reached the Yellowstone and undertook to swim it, the chances were decidedly against their succeeding.

Very sincerely,

C. A. Woodruff,

Brigadier-General, United States Army, Retired.

V.

So soon as this appendix as above was in type, I sent printed proofs of it to Generals Hughes, Woodruff, and Carrington, and to Colonel Godfrey for final revision and correction before the matter was plated.

In returning the proof, General Carrington and Colonel Godfrey both add further communications, which I insert below.

I also sent the same proof to Mrs. Elizabeth B. Custer, widow of General Custer, and to Mrs. John H. Maugham, his sister, with an expression of my willingness — nay, my earnest desire — to print any comment they or either of them might wish to make upon the question under discussion.

At Mrs. Custer's request I sent the appendix to Lieutenant-Colonel Jacob L. Greene, U. S. V., now president of the Connecticut Mutual Life Insurance Company, who was Custer's adjutant-general during the war and his life-long friend thereafter. His able defense of his old commander is printed as the last of this interesting series of historic documents.

Desiring that Custer, through his friends, may have the final word, I print it without comment, save to say that I fully join Colonel Greene in his admiration for the many brilliant qualities and achievements of his old commander.

General Carrington's Letter

Hyde Park, Mass., Sept. 25, 1904.

Dear Dr. Brady:

I appreciate the favor of reading the proof-sheets of the appendix to your papers upon the Custer massacre. When it occurred I was greatly shocked by an event so similar in its horrors to that of the Phil Kearney massacre, in 1866. A previous interview with General Custer came to mind, and I attended the sessions of the court of inquiry at Chicago, taking with me, for reference, a map which I had carefully prepared of that country, with the assistance of James Bridger, my chief guide, and his associates.

The evidence indicated that when Custer reached the "Little Big Horn" (so known upon that map) and sent Benteen up stream, with orders that, "if he saw any Indians, to give them hell," ordering Reno to follow the trail across the river and move down toward the Indian camps, while he moved down the right bank, detaching himself from the other commands, he practically cut the Indians off from retreat to the mountains, which was part of his special mission; but, in the flush of immediate battle, lost thought of the combined movement from the Big Horn, which had for its purpose the destruction of the entire Indian force by overwhelming and concentrated numbers.

Indeed, the court of inquiry did not so much discredit the conduct of Reno as reveal the fact that he faced a vastly superior force with no assurance that he could have immediate support from the other battalions, so vital in a sudden collision with desperate and hard-pressed enemies. The succeeding fight, on the defensive, protracted as it was, with no information of Custer's position, or possible support from him, was a grave commentary upon the whole affair.

The interview with General Custer referred to was in 1876, when, upon leaving the lecture platform of the Historical Society in New York, he made the remark, on our way to his hotel, "It will take another Phil Kearney massacre to bring Congress up to a generous support of the army." We spent several hours together, while he discussed his troubles with the authorities at Washington. He recalled the events of 1867, and felt that General Sherman had severely judged his operations on the Republican, but that the time was near when he might have an opportunity to vindicate himself, and that, "if he again had a chance he would accomplish it or die in the attempt." He was practically on a leave of absence, and its extension was not his choice. Colonel Smith was sick, and he claimed the right to command his regiment, since it had been ordered to report as part of General Terry's command.

The famous sutlership scandal was fully discussed, and here there has been confusion as to Custer's position. He had nothing to do with the popular complaint that Belknap was farming out sutlerships for personal emolument. Neither is it technically correct that the Secretary of War could make original appointments of the kind complained of. Post commanders, with their councils of officers, had both the selection of their sutler, and fixed the prices of articles to be sold. The Secretary simply issued the appointment thus designated, unless for good reasons declined, thereby requiring another selection by the officers. In cases of troops in campaign, or detached, or on distant service, the commanding General confirmed the officer's choice. Custer's position was manly, legal, and just; but his assertion of this right, so far as made, offended Belknap, at the expense of officers whose rights were overruled by non-military influence.

Custer was *not under charges* that would militate against his assignment to the rightful command of his regiment when ordered into field service. I did not hesitate to urge him to press his claim, but could not entertain the idea that he would go to Bismarck, or otherwise to make his claim in person, except through Washington Headquarters.

If ever a man had an incentive to dare odds with his regiment, this fearless fighter and rider, whose spirit reached the verge of frenzy in battle, was the man for the occasion.

Through all the papers cited by you, there runs the same subtle suggestion that he who, as an independent commander of aggressive cavalry in the Civil War, was almost expected to take into the field a large discretion as to his actions (whereby he had formerly achieved success) when confronted by the enemy, within striking distance in the Little Big Horn Valley, lost all sense of danger and all thought of prescribed details of action in the confidence that, somehow, the old Seventh could not be whipped by any savage force whatsoever!

I have always regarded Terry's general plan as well conceived, for Reno's prior scouting had almost assured the inevitable course of the Indian trail westward, and events confirmed Terry's judgment. General Hughes had served upon Terry's staff during the Civil War with credit, as well as captain in the 18th Infantry on the frontier, and his assurances that General Terry fully explained to Custer the reasons why Washington authorities distrusted his discretion and was more precise in giving him this detached command, cannot be impeached by an asserted affidavit that whispered hints, unheard by officers by his side, allowed him to be his own master in a matter where a combined movement of three commands was the prime factor in complete success.

Neither is there any doubt whatever that Custer's earnest plea, that he be trusted to fulfil the exact duties assigned to his command, secured not only the sympathy and confidence of Terry in his behalf, but that on that condition only did the Washington authorities authorize General Terry to vacate the order for his arrest because of going to his command without orders.

As already stated, Custer's confidence in the Seventh Cavalry was well deserved. It, with him, was a veritable thunderbolt in action; but it was not omnipotent. That over-confidence which dissolved its unity at the supreme crisis was fatal. Even then, a realized success of which Custer had no doubt, would have minimized the rashness of his dash and have largely condoned his fault.

Yours sincerely,

HENRY B. CARRINGTON.

COLONEL GODFREY'S FINAL REMARKS

I have no desire to pose as the special champion of General Custer, and it is still further from my desire to pose as inimical to General Terry. My only purpose is to demonstrate the truth, not only for this discussion, but for history.

This subject surely has gotten to the stage of academic discussion. I am not willing to admit that the phrase "he desires that you should conform to them (his views) unless," etc., conveys a direct, positive command which could not be more explicit. Nor do I admit that orders given by a commander, in which he uses the words "desires," "wishes," and equivalents, convey *positive* commands under all circumstances. In personal or social matters, such words convey the idea of what is wanted and what is expected; and in such matters the expressed wishes and desires are usually conveyed to personal friends, who loyally conform thereto, if not in letter, in spirit and in results. In such relations a commander does not want to use language that would appear dogmatic. I further admit that in personally giving orders a commander may accentuate the expression of his desires, wishes, etc., so as to leave no doubt about his intentions, and to convey positiveness thus expressed to his commands. When a commander gives written orders through official channels, the words "commands," "orders," and "directs," or the use of the imperative, leave little ambiguity or doubt as to what is ordered or intended.

Developments subsequent to the campaign or battle leave little doubt that General Terry had about him men or influences that were suspicious, inimical, or hostile to General Custer. I sincerely believe General Terry was too high minded knowingly to allow himself to be influenced by any sinister motive.

That the "instructions" give rise to this discussion shows they were vague. Was this vagueness intentional? General Terry was a lawyer. He was a soldier. As lawyer and soldier his trained mind should have weighed the words embodied in these instructions. Now read them: "It is of course *impossible** to give you any *definite instructions* in regard to this movement, and were it not *impossible* to do so, the Department Commander places too much confidence in your zeal, energy, and ability *to wish to impose upon you precise orders* which might hamper your action when nearly in contact with the enemy," and then goes on to indicate what he thinks should be done; or, in other words, indicates what he (Terry) himself would do if he found conditions as expressed. Custer evidently saw "sufficient reasons for departing from them" and did what a reasonable interpretation of the instructions contemplated, *made his own plans*.* I interpret the phrase "when so nearly in contact with the enemy" to refer to the immediate time or place (June 21, mouth of Rosebud) of writing it.†

* No italics in original. — E. S. Godfrey.

† This is interesting, and is the first suggestion I have met with that the phrase refers

As to the location of the Indians. Terry believed they were on the Little Big Horn;—we found them on that river about 15 miles above its forks with the Big Horn. Had the village been at the forks, the attack would have been delivered on the 25th of June, as the village would not have been located by Custer from the divide. It is possible the two columns might have joined in the attack. Now, suppose the village had been located 50 or 60 miles farther south, it would have still been within Terry's guess, but it would have been a far cry to Gibbons' column which, under the instructions, would have remained at the forks. It must be remembered that Custer would have had the Wolf Mountains (Rosebud Mountains on later maps) between him and the Little Big Horn had he ignored the trail and gone on southward up the Rosebud, as Custer's critics would have us believe were the intentions of the instructions.

General Woodruff would have him stop at the camp of June 24 and scout to locate the village, etc. Would that have complied with Woodruff's interpretation of the instructions? And from that position he says: "Custer was in a position to strike either one of the three last lines of flight (east, northeast or southeast), whereas if, after making the *forced* * night march with his fatigued animals, he had struck the Little Big Horn, and a reconnaissance had shown that the village had left the Little Big Horn going northeast, on June 24 he would have been two days' march behind them. That "forced" night march was about eight miles, and every mile made was in the direction to place us in the best position to intercept any flight to the northeast and east. Instead of being two days behind them, we would have met them almost "head on."

Of what practical use to send scouts through to Gibbon June 24?

There was no fresh or new positive information to send to him; Terry had "guessed" it all.

Now let us repeat the marches made: June 22, twelve miles; June 23, thirty-three miles; June 24, twenty-eight miles; June 25, eight miles to the bivouac; and ten miles to the divide, and then say fifteen miles to the village. That is to say, ninety-one miles up to noon June 25, when it was decided to attack, and one hundred and six miles in all four days. That doesn't indicate that we made *forced* marches.

to the position of Terry and Custer when the orders were prepared or delivered, and not to the time anticipated when Custer should meet the Sioux. I regret that I cannot agree with this interpretation. Still, it is possible that such an interpretation is certainly a point for Custer. — C. T. B.

* Italics mine. — E. S. GODFREY.

Woodruff further states that "he made that fatal night march with the *deliberate*(?) intention of trying to locate and strike the village before Gibbon could possibly get up." I say that statement is deliberately unfair, and contradicts the twice-told statement by Custer, that he did not intend to attack the village until the 26th, once before he knew the location of the village, the night of the 24th, and again when he called the officers together after the discovery at the divide.

Reno's position in the bottom, in the old river bed, was sheltered from fire from the hills by heavy timber, and was nearly a mile from the hills. I have never before heard that he was fired upon from those hills; but he was fired upon from the woods on the opposite side of the river. General Gibbon and I both thought the hills were too far away to give any effective fire. It must be remembered that the river bottom was heavily timbered for some distance above and below this position. This timber subsequently was cut for the construction of Fort Custer.

Lieutenant-Colonel Greene's Defense of Custer

Hartford, Conn., September 1, 1904.

My Dear Sir:

I have read with great interest your discussion of the question of General Custer's alleged disobedience of orders, both in the narrative of the Battle on the Little Big Horn and in the appendix to the volume, and upon which you have asked my comment.

For whatever bearing it may have upon the propriety of any comment of mine, let me say that General Custer was my intimate friend, and that his first act after receiving his appointment in the Civil War as a brigadier-general was to secure my appointment and detail to him as adjutant-general, which relation I held until his muster out of the volunteer service in 1866. I think no one knows better his quality as a soldier and as a man. I know his virtues and his defects, which were the defects of his virtues. He was a born soldier, and specifically a born cavalry man. The true end of warfare was to him not only a professional theory — it was an instinct. When he was set to destroy an enemy, he laid his hand on him as soon as possible, and never took it off. He knew the whole art of war. But its arts and its instruments and their correct professional handling were not in his eyes the end all of a soldier's career, to be satisfied with a technical performance. They were the means and the tools in the terms of which and by the use of which his distinct military genius apprehended and solved its practical and fateful problems. When he grappled his task it was to do it, not to go correctly through the proper motions to their technical limit, and then hold himself excused.

He was remarkable for his keenness and accuracy in observation, for his swift divination of the military significance of every element of a situation, for his ability to make an instant and sound decision, and then, for the instant, exhaustless energy with which he everlastingly drove home his attack. And the swiftness and relentless power of his stroke were great elements in the correctness of his decisions as well as in the success of his operations. He was wise and safe in undertaking that in which a man slower in observation, insight and decision, and slower and less insistent in action, would have judged wrongly and failed.

I knew Custer as a soldier when he was a brigade and division commander under Pleasanton and Sheridan, the successive commanders of the Cavalry Corps of the Army of the Potomac. Those who knew the estimate in which those great commanders held him — the tasks they committed to his soldierly intelligence and comprehension, his fidelity and skill — need no reminder that in nothing of all their dependence upon and confidence in him did he ever fail in letter or spirit. I know how absolutely loyal he was under the conflicting conditions which sometimes confront every subordinate charged with grave responsibilities, and which test the sense of duty to the utmost. He was true as steel. He was depended upon for great things because he was dependable.

In temperament he was sanguine and ardent. He loved his friends; he was impatient of every form of inefficiency and of pretense; he did not highly esteem mere professionalism; he was impulsive and sometimes abrupt in manner, but kind of heart; he was sensitive only to unjust criticism; he despised intrigue, chicane and all meanness; he was independent in opinion and judgment, and frank in their expression; he was open in opposition, and fair to an enemy.

And it goes without saying that such a man had enemies — men who were envious of his abilities, his achievements and his fame; men whom he never sought to placate, and who sought envy's balm in detraction and hatred; men who could not measure him or be fair to him, but men who in a pinch would have turned to him with unhesitating trust, whether in his ability or his soldierly faith.

Did this man, this soldier, whose service throughout the Civil War and a long career of frontier warfare was for eighteen years unequaled for efficiency and brilliancy within the range of its opportunities and responsibilities, who never failed his commanders, who never disobeyed an order, nor disappointed an expectation, nor deceived a friend — did this man, at the last, deny his whole life history, his whole mental and moral habit, his whole character, and wilfully disobey an understood order, or fail of its right execution according to his best judgment, within the limits of his ability under the conditions of the event; and, what is worse — and this is what his detractors charge — did he not only disobey, but did he from the inception of the enterprise plan to disobey — to deceive his commander who trusted him, in order that he might get the opportunity to disobey ?

To any man who knew Custer, except those who for any reason hold a brief against him, not only is the charge of premeditated, deliberate disobedience absurd, but it is a foul outrage on one of the memories that will never fail of inspiration while an American army carries and defends an American flag.

In one of Mrs. Custer's letters to me, narrating what took place during the days of preparation for the General's departure, she wrote:

"A day before the expedition started, General Terry was in our house alone with Autie (the General's pet name). A.'s thoughts were calm, deliberate, and solemn. He had been terribly hurt in Washington. General Terry had applied for him to command the expedition. He was returned to his regiment because General Terry had applied for him. I know that he (Custer) felt tenderly and affectionately toward him. On that day he hunted me out in the house and brought me into the living-room, not telling me why. He shut the door, and very seriously and impressively said: 'General Terry, a man usually means what he says when he brings his wife to listen to his statements. I want to say that reports are circulating that I do not want to go out to the campaign under you.' (I supposed that he meant, having been given the command before, he was unwilling to be a subordinate.) 'But I want you to know that I do want to go and serve under you, not only that I value you as a soldier, but as a friend and a man.' The exact words were the strongest kind of a declaration that he wished him to know he wanted to serve under him."

That was Custer all over. And to any one who knew him — to any one who can form a reasonable conception of the kind of a man he must needs have been to have done for eighteen years what he had done and as he had done it, and won the place and fame he had won — that statement ends debate. Whatever of chagrin, disappointment, or irritation he may have felt before, however unadvisedly the sore-hearted, high-spirited man may have spoken with his lips when all was undetermined, and his part and responsibility had not been assigned, this true soldier, knowing the gossip of the camp, conscious possibly it was not wholly without cause, however exaggerated, but facing now his known duty and touched by the confidence of his superior as Custer never failed to be touched, could not part from his commander with a possible shadow resting between them. He knew the speech of men might have carried to Terry's mind the suggestion of a doubt. And yet Terry had trusted him. He could not bear to part without letting General Terry know that he was right to trust him. That statement to Terry was a recognition of whatever folly of words he might before have committed in his grief and anger; it was an open purging of an upright soldier's soul as an act honorably due alike to superior and subordinate; it was, under the circumstances, the instinctive response of a true man to the confidence of one who had committed to him a trust involving the honor and fame of both. Disobedience, whether basely premeditated, or with equal baseness undertaken upon after-deliberation, is inconceivable, unless one imputes to Custer a character void of every soldierly and manly quality. With such an one discussion would be useless.

Upon the discussion itself, which is presented in the narrative and in the appendix, I have little to say. In the opening paragraph of the appendix, you say: "I presume the problem . . . will never be authoritatively settled, and that men will continue to differ upon these questions until the end of time."

In other words, the charge of disobedience can never be proved. The proof does not exist. The evidence in the case forever lacks the principal witness whose one and only definite order was to take his regiment and go "in pursuit of the Indians whose trail was discovered by Major Reno a few days since." They were the objective; they were to be located and their escape prevented. That was Custer's task. All the details were left, and necessarily left, to his discretion. All else in the order of June 22d conveys merely the "views" of the commander to be followed "unless you should see sufficient reasons for departing from them." The argument that Custer disobeyed this order seems to resolve itself into two main forms. One is trying to read into the order a precision and a peremptory character which are not there and which no ingenuity can put there, and to empty it of a discretion which is there and is absolute; the other is in assuming or asserting that Custer departed from General Terry's views without "sufficient reasons." And this line of argument rests in part upon the imputation to Custer of a motive and intent which was evil throughout, and in part upon what his critic, in the light of later knowledge and the vain regrets of hindsight, thinks he ought to have done, and all in utter ignorance of Custer's own views of the conditions in which, when he met them, he was to find his own reasons for whatever he did or did not do. Under that order, it was Custer's views of the conditions when they confronted him that were to govern his actions, whether they contravened General Terry's views or not. If in the presence of the actual conditions, in the light of his great experience and knowledge in handling Indians, he deemed it wise to follow the trail, knowing it would reach them, and deeming that so to locate them would be the best way to prevent their escape, then he obeyed that order just as exactly as if, thinking otherwise, he had gone scouting southward where they were not, and neither Terry nor he expected them to be.

To charge disobedience is to say that he wilfully and with a wrong motive and intent did that which his own military judgment forbade; for it was his own military judgment, right or wrong, that was to govern his own actions under the terms of that order. The quality of his judgment does not touch the question of obedience. If he disobeyed that order, it was by going contrary to his own judgment. That was the only way he could disobey it. If men differ as to whether he did that, they will differ.

Respectfully yours,

Jacob L. Greene.

VI.

To sum up, I suggest this as a possible line of investigations by which the student may determine the question for himself:

First. Were Terry's written orders definite and explicit?

Second. Were they intelligent orders capable of execution?

Third. Did these orders admit of more than one meaning?

Fourth. What are the various meanings, if more than one?

Fifth. Did Custer carry them out in any of their meanings?

Sixth. Did Custer depart from them?

Seventh. If so, how far?

Eighth. Such being the case, was he justified in so departing by the exigencies of the situation?

Ninth. Were the consequences of such a departure serious?

Tenth. Did Custer receive verbal orders from Terry at the last moment?

Eleventh. If so, when?
Twelfth. Did these verbal orders supersede the written orders?

In closing, I repeat that I should be glad to be convinced that I have erred in my conclusions; and that if any one can convince me that Custer did not disobey, or that in doing so he was justified in his disobedience, I shall make the fullest public amends for my expression of opinion that he did and that he was not.*

* At the risk of tiring the reader, but because I am sensitive in the matter and anxious not to be misunderstood, I append here a letter written by me to a sister of General Custer, who had expressed the hope that I would not take the position that he disobeyed his orders

July 13th, 1904.

My Dear Madam:

I have received and read and reread your letter of the 12th inst. That letter and the thought of Mrs. Custer, whose character, in common with all Americans, I respect and admire, taken in connection with the position which my conscience, much against my will, has compelled me to assume, has filled me with deep regret.

Having read thus far, you will undoubtedly divine that I am compelled to say that I believe General Custer did disobey his orders. I have nowhere stated that I consider him guilty of rashness. I have also made it plain, I think, that even though he did disobey his orders, the ultimate annihilation of his battalion was due to the cowardice or incapacity of Major Reno.

I remember to have seen General Custer when I was a boy in Kansas. My father, who was a veteran of the Civil War, had a great admiration for him. I was present when the bodies of the officers of the Seventh Cavalry were brought back for reinterment at Fort Leavenworth. My wife, a Southern woman, is a cousin of the late General Dod Ramseur, who was General Custer's intimate friend. The family have never forgotten General Custer's kindness when Ramseur was killed.

I did, and still have, a warm admiration for the brilliant and soldierly qualities of General Custer. He was, and is, my beau ideal of a cavalry soldier. When I began to write these articles, I would not hear the charge that he had disobeyed orders. But I have been compelled by my investigations to take that position. I cannot tell you how painful it has been to me, and it is, to come to this conclusion. I have thought long and deeply over the matter.

Of course I read General Hughes' now famous article. I did not, however, allow that article alone to determine me; but I carefully considered every account. I examined every discussion which I could find. Not only that, I corresponded with a number of officers, among them being Lieutenant-General Miles, Major-General Hughes, Brigadier-General C. A. Woodruff, Brigadier-General Carrington, and Colonel Godfrey. The remarks of these officers were submitted to one another. Their statements were weighed and digested with the utmost care by me. I could come to no other conclusion than that I have arrived at.

As an Appendix I have inserted in full my correspondence with different officers concerning the matter. I have been glad to print all that Colonel Godfrey, who has indeed been a powerful advocate in opposition to my views, has written. I have called attention to one significant fact which, in my opinion, would fully clear General Custer from the charge of disobedience. That is the affidavit of an alleged witness to the last conversation between Terry and General Custer.

General Miles refers to this affidavit in his book, "Personal Recollections of General Nelson A. Miles." President E. Benjamin Andrews also refers to it in his book, "The United States in our Own Time." I wrote to President Andrews, who gave Miles as his authority. I wrote to General Miles *three times*, registering the last letter, asking him to substantiate the affidavit, the existence of which was doubted by many army officers. General Miles has made no reply. I take it for granted, therefore, that *he cannot substantiate the affidavit.*

I have said frankly that if he can prove this affidavit and establish the credibility of the affiant, I will make public amends in the most ample manner for having said General Custer disobeyed his orders. I have said that if *anybody* can convince me that I have been wrong in my concluison ; if any evidence can be produced which will establish the contrary, I shall be most happy to retract what I have said in any possible way that may be suggested to me.

I beg you to believe that I have written in no spirit of animosity to General Custer. My real feelings for General Custer can easily be seen from my article on the Battle of the Washitá, to which you have referred. It would be most agreeable to me if you would forward this letter to Mrs. Custer.

Again deploring the unfortunate conclusion I am conscientiously compelled to arrive at, and regretting more than I can express that I must give pain to Mrs. Custer, to you, and to the friends of General Custer, I am,

Yours sincerely,

CYRUS TOWNSEND BRADY.

If this Appendix shall cause any light to be thrown on the affidavit so often referred to, it will serve an excellent purpose; for, I say again, I shall consider the establishment of that affidavit as settling the question.

The subject is now left with the student. Perhaps I cannot more fitly close the discussion than by this quotation from the confidential report of General Terry by General Sheridan, dated July 2, 1875:

"I do not tell you this to cast any reflection on Custer, for whatever errors he may have committed he has paid the penalty, and you cannot regret his loss more than I do; but I felt that our plan must have been successful had it been carried out, and I desire you to know the facts."

APPENDIX B

Further Light on the Conduct of Major Reno

AFTER the publication of the Custer article censuring Reno, my attention was called to the following editorial, which appeared in the *Northwestern Christian Advocate*, of September 7, 1904:

Why General Custer Perished

General George A. Custer was and will always be regarded as one of the most brilliant officers of the United States Army. His career abounds in romantic interest; and his death, together with that of every officer and soldier fighting with him, was one of the most tragic and memorable incidents in Indian warfare. The story of Custer's last fight with the Indians, which took place on the Little Big Horn River in the summer of 1876, is graphically described by Cyrus Townsend Brady. It is not our purpose to relate the story of the battle, but to call attention to the real cause of Major Reno's conduct, which resulted in Custer's defeat and death.

After describing the movements by which Custer distributed his force, and the task assigned to Major Reno, who displayed remarkable indecision and errors of judgment, which would have been inexcusable even in an inexperienced young officer, and caused Reno to retreat instead of vigorously attacking the Indians, Mr. Brady says:

"His [Reno's] second position was admirable for defense. Sheltered by the trees, with his flanks and rear protected by the river, he could have held the place indefinitely. He had not, however, been detailed to defend or hold any position, but to make a swift, dashing attack; and after a few moments of the feeblest kind of advance, he found himself thrown upon the defensive. Such a result would break up the most promising plan. It certainly broke up Custer's.

"It is a painful thing to accuse an army officer of misconduct, but I have taken the opinion of a number of army officers on the subject, and every one of them considers Reno culpable in a high degree. One, at least, has not hesitated to make known his opinion in the most public way. I am loath to believe that Major Reno was a coward; *but he certainly lost his head, and when he lost his head he lost Custer.* His indecision was pitiful. Although he had suffered practically no loss and had no reason to be alarmed, he was in a state of painful uncertainty as to what he should do next. The soldier — like the woman — who hesitates in an emergency which demands instant decision is lost. . . .

"There had, as yet, been no panic, and under a different officer there would have been none; but it is on record that Reno gave an order for the men to mount and retreat to the bluffs. Before he could be obeyed he countermanded this order. Then the order was given again, but in such a way that nobody, save those immediately around him, heard it because of the din of the battle then raging in a sort of aimless way all along the line, and no attempt was made to obey it. It was then repeated for the third time. Finally, as those farthest away saw those nearest the flurried commander mounting and evidently preparing to leave, the orders were gradually communicated throughout the battalion and nearly the whole mass got ready to leave. Eventually they broke out of the timber in a disorderly column of fours, striving to return to the ford they had crossed when they had entered the valley.

"Reno calls this a charge, and he led it! He was so excited that, after firing his pistols at the Indians, who came valiantly after the fleeing soldiers, he threw them away. The pressure of the Indians upon the right of the men inclined them to the left, away from the ford. In fact, they were swept into a confused mass and driven toward the river. All semblance of organization was lost in the mad rush for safety. The troops had degenerated into a mob."

Major Reno was not a coward, as many believe. His career in the army during the Civil War and his promotion for gallant and meritorious services at Kelley's Ford, March 17, 1863, and at the battle of Cedar Creek, October 19, 1864, are evidence of his courage. What, then, was the explanation of his conduct at the Battle of the Little Big Horn? Dr. Brady does not give it. Perhaps he does not know. But Major Reno himself told the late Rev. Dr. Arthur Edwards, then editor of the *Northwestern*, that his strange actions were due to the fact that HE WAS DRUNK. Reno's conduct in that battle lost him many of his military friends. To Arthur Edwards, who knew him well, and continued his faithful friend, Major Reno often unburdened his heart, and on one occasion in deep sorrow said that his strange actions were due to drink, and drink ultimately caused his downfall. His action at the Battle of the Little Big Horn was cited as one instance of the result of his use of intoxicating liquor. Liquor finally caused his expulsion from the army in disgrace. In 1880 he was found guilty, by a general court-martial, of conduct unbecoming an officer and a gentleman. While in an intoxicated condition he engaged in a brawl in a public billiard saloon, in which he assaulted another officer, destroying property and otherwise conducted himself disgracefully. For this offense the court sentenced him to be dismissed from the army.

It had occurred to me that probably the explanation of Reno's conduct lay in the fact that he might have been intoxicated. I asked Colonel Godfrey if he thought so, and his reply has been noted above in Appendix A.

After reading the article in the *Advocate*, I wrote to the editor, Dr. David D. Thompson, asking for further evidence of the statement quoted. Here follows his letter:

Chicago, September 30, 1904.

My Dear Sir:

Doctor Arthur Edwards, the former editor of the *Northwestern Christian Advocate*, was chaplain in the army during the Civil War. He was a soldier by instinct, and kept up his interest in military and naval affairs and his acquaintance with army and naval officers during all his life. In the army he won the confidence of his fellow-officers by his character and moral courage.

He was requested, by a number of officers, to wait upon General Hooker, then in command of the army, and express to him the great anxiety felt by the officers over his intemperate habits. Doctor Edwards waited upon General Hooker, and told him what the officers had requested him to say. He did it in so manly and delicate a way that General Hooker thanked him, and told him the army would not again have occasion to fear ill results because of his habits.

The story of this incident came to the knowledge of Mr. H. I. Cleveland, an editorial writer on the Chicago *Herald*, who published it several years ago over his own name in that paper. I had never before heard the story from Doctor Edwards, and when I saw Mr. Cleveland's article I asked Doctor Edwards about it. He related the story to me, and, after doing so, told the story of Reno as I give it briefly in the *Northwestern*.

From all that I can learn of Reno, the feeling in the army against him was not due to his drinking habits, but to his conduct in his relations with others. Doctor Edwards told me that Reno told him that all of his trouble in his contact with his fellow-officers was due, primarily, to his drinking habits, which had undermined his moral character. Doctor Edwards knew Reno very well, and told me he believed that drinking was, as Reno himself stated, the cause of all his trouble. He had known him in the army during the Civil War, and spoke highly of his character as a soldier at that time.

Yours sincerely,

D. D. THOMPSON.

P. S.—Doctor Edwards intended at some time to publish this story, but died in April, 1901, before doing so.

II.

As I have always been most willing and anxious to give the accused a hearing in every case, it gives me great pleasure to insert here a letter recently received from Mr. William E. Morris, an attorney, who is also an alderman of Greater New York. In this letter will be found a spirited defense of Major Reno, with interesting details of his fight. Although Mr. Morris dissents from many of my conclusions, and differs radically from the printed accounts of Colonel Godfrey and others, I am glad to place the other side before my readers. I only regret that this paper was received too late to be included in the body of the book.

Haven, Maine,
September 21, 1904.

Dear Sir:

I have read your article entitled "War with the Sioux," and as a survivor of Reno's Battalion desire to enter an earnest protest against the many incorrect statements of alleged facts.

Col. Reno was cruelly libeled while he was alive, and took his medicine manfully, knowing that he had the respect of every officer and enlisted man who served under him on the 25th and 26th days of June, 1876.

The 7th Cavalry had no use for cowards, and had Reno showed the white feather, he would have been damned by every member of his command.

As a matter of fact, we revere his memory as that of a brave and gallant officer, who, through circumstances over which he had no control, was blamed by the public, who had no personal knowledge of the facts for the result of the Battle of the Little Horn.

It is quite evident to me that you have never interviewed a single member of Reno's Battalion, to wit: Troops "A," "G," and "M," for if you had you would not misstate the facts, as I assume that you intend to be fair, and would not intentionally mislead the public mind.*

*I have been in communication with a number of persons who belonged to this battalion.—C. T. B.

I was a member of Capt. Thomas H. French's Troop "M," 7th U. S. Cavalry, and I submit the following as a concise statement of the facts:

We lost sight of Custer, whose command was on our right, at least thirty minutes before we crossed the Little Horn River.

We saw a party of about one hundred Indians before we reached the river; we pursued them across the Little Horn and down the valley. As soon as we forded, Reno gave the command, "Left into line, gallop — forward, guide, center," and away we went faster than I had ever ridden before. The Indians rode as fast as they could, and the battalion in line of battle after them. A body of at least two thousand came up the valley to meet the one hundred or more we were pursuing. They immediately made a flank movement to our left and a stand, opened a galling fire, causing some of our horses to become unmanageable. John R. Meyer's horse carried him down the valley through the Indians, some of whom chased him two or three miles over the hills and back to ford. He escaped with a gun-shot wound in the neck. Rutten's horse also ran away, but he succeeded in making a circle before reaching the Indians, and received only a gun-shot wound in the shoulder. We were then abreast the timber; to continue the charge down the valley meant (to the mind of every one) immediate destruction of the battalion, which consisted of about one hundred and twenty men (the old guard, of ten men from each troop, being with the packs).

Reno, very properly, gave the command "Battalion halt — prepare to fight on foot — dismount!" He directed French to send ten men from the right of his troop to skirmish the woods, before the "numbers four" proceeded there with the horses. We immediately deployed as skirmishers and opened fire. The odds were at least thirty to one, as our line with the fours out did not exceed seven officers and ninety men. We had, however, a few Indian scouts and civilians. We had entire confidence in our officers and in ourselves, and went to work smiling and as cool as if we were at target practice. In less time than it takes to relate it, the Indians were on three sides of us. We were ordered to lie down, and every man that I could see, except Reno and French, were fighting lying down. Reno walked along the line giving instructions to the men, while French was calling his men's attention to his own marksmanship with an infantry long-tom that he carried.

While in this position, the man next on my right, Sergeant O'Hara, was killed. The smoke obscured the line, but bullets were taking effect all along it. We were perfectly cool, determined, and doing good execution and expected to hear Custer attack. We had been fighting lying down about fifteen minutes when one of our men came from the timber and reported that they were killing our horses in the rear. Every troop had, at this time, suffered loss and the enemy was closing in, despite our steady and deadly fire. Reno then made his only error; he gave the command, "Retreat to your horses, men!" French immediately corrected the mistake with the command, "Steady, men — fall back, slowly; face the enemy, and continue your fire." "M" troop fell back slowly and in perfect order, held the Indians in check until "A" and "G" had mounted. Several of their horses had been shot, and their riders, consequently, very much disturbed.

"M" Troop left Sergeant O'Hara and Private Smith on the skirmish line. Isaiah, the colored interpreter of Fort Rice, Bloody Knife, the Chief of the Rees Scouts, and a civilian also remained. Lawrence was hit in the stomach when about to mount. I went to his relief, which caused me to be the last man to leave the timber, with the command, with the exception of Lieutenant Hare, who passed me in the bottom. Sergeant Charles White was wounded in the arm and his horse killed. He was left in the woods, as was also "Big Fritz," a Norwegian, whose surname I do not remember, but whose horse was killed. "A" and "G" had men left in the timber also, and they all reached the command on the hill during the night with De Rudio, or about the same time.

I give more details in regard to "M" than the other troops, because of a personal acquaintance with each member. Corporal Scollen and Private Sommers fell in the charge from the timber to the ford. It was a charge and not a retreat, and it was led by Reno. Every man that I saw used his revolver at close range. I was at least twenty yards behind the rear of the command. The Indians closed in, so I was compelled to jump my horse off the bank, at least fifty yards below the ford, and while in the river had an excellent view of the struggle. It was hand to hand, and McIntosh was certainly there at the ford and sold his life as dearly as he possibly could. When I reached the cut in bank, I found Turley and Rye mounted and Lieutenant Hodgson wounded and dismounted. He was waist-deep in the water. He grasped my off stirrup strap with both hands. Rye let Turley go ahead through the cut, and he was killed as he reached the top; Rye followed without receiving a scratch. The lieutenant held onto my stirrup for two or three seconds, and was dragged out of the water. He was hit again, and let go as my horse plunged up the cut. Sergeant Criswell may have assisted him out of the water, but if he did he went back into it again. To say that any man could or did ride back down that cut is to suggest, to my mind, the impossible. Upon reaching the level above the cut I dismounted and led my horse as fast as possible up the bluff, and overtook Tinker, Bill Meyer and Gordon about half way up the bluff. We stopped a moment to rest. The bodies of the fallen soldiers were plainly visible. They marked the skirmish line and the line of the charge from the timber to the ford, and were in the river and at the top of the cut. At this instant a shower of lead sent Meyer and Gordon to the happy hunting-ground, and a fifty caliber passed through the left breast of your humble servant. Our horses were also hit. I continued up the hill alone and joined the command; was then assisted to the improvised hospital.

Reno at this time had lost, in killed, wounded, and left dismounted in the woods, over 30 per cent. of his battalion (there were over ten left in the woods). Lieutenant Hare was particularly conspicuous, and distinguished himself by his cool and determined manner when he ordered the men to fall in at the top of the hill, and whatever demoralization there was, was immediately dispelled by that courageous young Texan. Benteen, arriving about an hour later, came up as slow as though he were going to a funeral. By this statement I do not desire to reflect in any way upon him; he was simply in no hurry; and Müller, of his troop, who occupied an adjoining cot to mine in the hospital at Fort Abraham Lincoln, told me that they walked all the way, and that they heard the heavy firing while they were watering their horses.

Benteen was, unquestionably, the bravest man I ever met. He held the Indians in absolute contempt, and was a walking target from the time he became engaged until the end of the fight at sundown on the 26th. He took absolute charge of one side of the hill, and you may rest assured that he did not bother Reno for permission of any kind. He was in supreme command of that side of the hill, and seemed to enjoy walking along the line where the bullets were the thickest. His troop, "H," did not dig rifle-pits during the night of the 25th, as the other troops did, and in the morning their casualties were increased on that account. He ordered "M" out of their pits to reinforce his troop. There was some dissatisfaction at the order, as the men believed that the necessity was due solely to the neglect of "H," in digging pits. They obeyed, however, and assisted Benteen in his famous charge.

It was rumored, subsequently, that French recommended his First Sergeant, John Ryan, a sharpshooter, and some other men for medals, and that Benteen refused to indorse the recommendation as to Ryan, because he failed immediately to order the men out of their pits at his end of the line at his (Benteen's) order. It was claimed that French thereupon withdrew his list. Ryan was in charge of the ten men that Reno sent to skirmish the woods.

I was very much amused to learn, from your article, that Windolph received a medal. I remember him as the tailor of "H" troop, and have a distinct recollection of his coming into the field-hospital, bent almost double and asking for treatment for a wound which, his appearance would suggest, was a mortal one, but which the surgeon found, on removing his trousers, to be only a burn. The surgeon ordered him back to the line amid a shout of laughter from the wounded men. Mike Madden of "K" lost his leg, and Tanner of "M" his life, in the dash for the water for the wounded. I hope Madden received a medal.

In view of the conflict between the foregoing and the statements contained in your article, I ask you to investigate the matter further, with a view to correcting the false impression that your readers must have concerning Reno and his command. In conclusion, I ask you "how, in God's name," you could expect Reno, with one hundred and twenty men, to ride through upwards of three thousand armed Sioux, and then be of assistance to Custer or any one else? I say we were sent into that valley and caught in an ambush like rats in a trap. That if we had remained ten minutes longer, there would not have been one left to tell the tale. That the much abused Reno did charge out of the timber, and that we who survive owe our lives to that identical charge which he led. We, at least, give him credit for saving what he did of his command. I am, sir,

Very respectfully,
Wm. E. Morris,
Late private Troop "M," Seventh U. S. Cavalry.

INDEX

INDEX

I

J

N